VETERINARY FIELD STATION LIBRARY
013872323

THE UNIVERSITY OF LIVERPOOL
VETERINARY FIELD STATION LIBRARY
WITHDRAWN FROM STOCK

Histological Atlas of the Laboratory Mouse

AF749764

Histological Atlas of the Laboratory Mouse

William D. Gude
Oak Ridge National Laboratory
Oak Ridge, Tennessee

Gerald E. Cosgrove
San Diego Zoological Society
San Diego, California

and

Gerald P. Hirsch
Veterans Administration Wadsworth Hospital Center
Los Angeles, California

PLENUM PRESS • NEW YORK AND LONDON

Library of Congress Cataloging in Publication Data

Gude, William D.
Histological atlas of the laboratory mouse.

Bibliography: p.
Includes index.
1. Mice—Cytology—Atlases. 2. Histology—Atlases. I. Cosgrove, Gerald E., . II. Hirsch, Gerald P., 1939— . III. Title.

QL813.M55G8 599.32'33 81-8708
ISBN 0-306-40686-1 AACR2

© 1982 Plenum Press, New York
A Division of Plenum Publishing Corporation
233 Spring Street, New York, N.Y. 10013

All rights reserved

No part of this book may be reproduced, stored in a retrieval system, or transmitted, in any form or by any means, electronic, mechanical, photocopying, microfilming, recording, or otherwise, without written permission from the Publisher

Printed in the United States of America

Foreword

The Biology Division of the Oak Ridge National Laboratory was organized in 1946 for the purpose of studying the immediate and long-term implications of man's exposure to ionizing radiation. The program that developed concentrated on the basic mechanism of the effects in biological organisms from the genetic, biochemical, biophysical, and molecular biophysical points of view.

Most of its activities at the beginning concentrated on nonmammalian work (bacteria, fungi, *Drosophila,* plants, etc.) since no facilities to perform mammalian studies were available at that time. It became most obvious that specimens more closely related to mammalian tissue would likely yield more conclusive data to extrapolate these effects upon man.

At Oak Ridge, the first study was the Mammalian Genetics Study under William L. Russell designed to investigate the genetic changes produced by exposure to ionizing radiation and the implication for man. A couple of years later, this was followed by the study of somatic cell effects through the efforts of Jacob Furth, later by Arthur Upton, and finally by John Storer. In all of these assays, we concentrated on work with the mouse since more was known about its genetics than about the genetics of any other mammal. Studies could be conducted with very large numbers of mice, and mice proved to be especially suitable for cancer induction studies.

As this work progressed, we became convinced that a strong histology department was needed to prepare the tissues in a uniform manner and also to examine and interpret them. With the support of Dr. Furth at that time, we secured the services of William D. Gude, who organized this section of the Biology Division and whose dedicated management developed it into a central information source for histology work, not only for our Biology Division but also for this area of Tennessee, thus establishing its excellent reputation.

I am most pleased to see that Mr. Gude has assembled this work into a detailed atlas of the laboratory mouse. As essential as such information is to basic mammalian research, such a volume of complete data and illustrations is not always available in a single source, and I believe this volume will serve as an important handbook for researchers utilizing these species.

Alexander Hollaender

Associated Universities, Inc.
Council for Research Planning in Biological Sciences
1717 Massachusetts Avenue, N.W.
Washington, D.C. 20036

Acknowledgments

We are most appreciative of the encouragement by Dr. Alexander Hollaender, Associated Universities, Washington, D.C., to undertake this work and of his efforts in helping to arrange the publication of this book. Dr. Wanda Haschek of the Biology Division, Oak Ridge National Laboratory, has been most helpful in offering valuable suggestions concerning the selection of chapter headings and the clarification of histological descriptions. Our thanks go also to other members of the Biology Division for their suggestions and help, including Dr. Neal Clapp, Dr. E. F. Oakberg, Dr. Ray Popp, Dr. Kowetha Davidson, and Dr. J. M. Holland. Dr. Conrad Richter, Medical Division, Oak Ridge Associated Universities, kindly participated in the critical selection of photomicrographs, offering many suggestions for improvements. Dr. Desmond Doyle, University of Tennessee Veterinary School, helped in identifying specific structures of the male reproductive system. Gene Watkin of the Information Section, Fusion Energy Division, provided us with several ideas for improving the organization of the book and for planning the arrangement of colored prints.

We owe many thanks to Joyce Feezell and Jimmy Wesley for preparing excellent paraffin sections of the tissues and organs. Jimmy also prepared high-quality thin sections of methacrylate-embedded tissues. The photomicrographs were taken through a Zeiss Ultraphot II microscope using Kodak Professional Ektachrome Film 6118. The color film was processed by Color Technique, Inc., Chicago.

We want to express our gratitude to Mrs. Charlotte Rains for her diligence and patience in preparing the many typewritten copies of all sections of the manuscript.

We are grateful to Katharine Hummel for granting permission to copy descriptive material from the microscopic anatomy sections of Chapter 13, "Anatomy," of *Biology of the Laboratory Mouse*, 2nd ed., Blakiston Division, McGraw-Hill Book Co., New York, 1966.

For assistance with proofreading we are indebted to David A. Hambright and Craig Whitmire of the Technical Publications Department, Information Division, Oak Ridge National Laboratory.

The cover of the book was designed by Nancy Harrell Lockwood, Graphic Arts Department, Oak Ridge National Laboratory.

Research reported herein was sponsored by the Office of Health and Environmental Research, U.S. Department of Energy, under contract W-7405-eng-26 with the Union Carbide Corporation, and supported in part by U.S. Department of Energy contract EY-76-C-02-0016 with Associated Universities, Inc., Brookhaven National Laboratory.

W. D. G.
G. E. C.
G. P. H.

Contents

Histological Atlas of the Laboratory Mouse

Introduction

The purpose of this book is to provide investigators and students with an atlas of tissues of mice, commonly used experimental animals in laboratory studies. Many strains of mice are available for research and clinical laboratory studies, but histological differences between strains are usually minor. We have taken samples from six strains presently used in the Biology Division, Oak Ridge National Laboratory: BALB/c, RF, C3H, C57BL, $BC3F_1$/Cum, and $B6D2F_1$/Jax. Both males and females between the ages of 3 and 10 months were used. They were killed with ether and the dissected tissues were placed in an appropriate fixative immediately upon removal from the mouse. The choice of fixative depended on the stain to follow and included 10% neutral buffered formalin, Bouin's fixative, Zenker-formol (Helly's fluid), formal-mercuric chloride, and Williams's fixative (formulas given in the Fixatives section). Many of the sections pictured in this book were stained with hematoxylin and eosin (H & E). In addition, various special stains were used to illustrate specific structures or features. For example, mucus present in goblet cells of the intestine was stained brilliant red with periodic acid–Schiff reagent (PAS). Neurons and nerve ganglia stained black or dark brown with silver, as in the Bodian stain, are outstanding against a colorless background. Procedures for obtaining the color combinations of many other tissues are given in the Appendix for those who wish to reproduce them for their own investigative research.

With the development of embedding materials other than paraffin (methacrylate and epon) and the availability of microtomes that section tissues 1–2 μm thick, it is now practical to obtain very thin sections of cells in which fine cytoplasmic and nuclear structures can be identified that would be very difficult to see in a 5-μm-thick section.

We believe this atlas may also be used as a supplement to standard histology texts. For more detailed descriptions of tissues and organs characteristic of the mouse see *Biology of the Laboratory Mouse* (see Recommended Reading).

Nervous System
(Figures 1–19)

Irritability and conductivity are two characteristics of the nervous system that require specialized cells and tissues. Anatomically, the nervous system is divided into the central nervous system (CNS), including the brain and spinal cord, and the peripheral nervous system (PNS), which includes all other nerve elements in the organism (peripheral nerves, ganglia, etc.). Specialized cells that do not conduct nerve impulses but contribute supporting tissue to the nervous system comprise the neuroglia. Nervous tissue also may be divided into gray and white matter. Nerve cell bodies, dendrites, and unmyelinated portions of axis cylinders comprise gray matter. White matter consists primarily of myelinated nerve fibers, which give it a glistening white appearance. Neuroglial cells are present in both gray and white matter.

The functional, impulse-conducting unit of nervous tissue is the neuron. Its basic structure consists of a large globular nucleus within which are a few chromatin granules and usually one large nucleolus. The cytoplasm or perikaryon surrounding the nucleus contains abundant, irregular masses of basophilic material rich in ribonucleic acid also known as Nissl or tigroid substance. Following silver staining numerous fine threads of neurofibrils may be identified in the cytoplasm.

All neurons have one process extending from the cell body at the axon hillock, the axon or axis cylinder. Impulses travel away from the cell body down the axons, which decrease in diameter as the distance from the cell body increases. Collateral branches may be given off at intervals. Other processes, dendrites, are direct extensions of the cell body and they connect with axons of functionally related neurons at synapses or junctions. Impulses travel along dendrites toward the cell body from the synapsis.

The axis cylinder in peripheral nerves is covered by a neurolemma or Schwann's sheath. In nerves having a myelin sheath covering the axon the neurolemma forms the outermost covering. Nuclei of Schwann cells may be identified alongside the axis cylinder. A node of Ranvier occurs at the junction of two merging segments of myelin sheath.

Neurons may be further categorized by the number of processes extending from the cell body. Those having only one axon are unipolar neurons. Bipolar neurons have an axon and one dendrite extending from the cell body at opposite poles; multipolar neurons are determined by the number and arrangement of dendrites. Ganglia are the chief paths of communication between the CNS and outer ganglionic plexuses. Their nerve cell bodies are usually large and spherical with a predominant nucleolus and abundant chromatin in the cytoplasm.

Neuroglia or specialized interstitial tissue of the nervous system includes ependyma, a type of cuboidal or low columnar epithelium that lines the ventricles of the brain and spinal canal, and satellite cells of peripheral ganglia. Three types of neuroglia are identified: astrocytes (fibrous and protoplasmic), oliogodendroglia, and microglia. Protoplasmic astrocytes, which are found chiefly in gray matter, exhibit processes, many of which are attached to blood vessel walls and to the pia mater. Fibrous astrocytes, more commonly found in white matter of the brain between nerve fibers, have a larger elongated nucleus than the protoplasmic type, but they also send out processes that attach to blood vessel walls.

Oliogodendroglia resemble astrocytes but are much smaller cells and have only a few processes with few branches. They are associated with supporting nerve fibers rather than blood vessels, and are called "satellite" cells when found adjacent to nerve cell bodies. Microglia have a small nucleus and a small amount of cytoplasm. They are scattered throughout the brain and spinal cord.

The meninges, connective tissue coverings of the brain and spinal cord, include (1) the outermost dura mater, a dense and tough material next to the bone, (2) the arachnoid, a thin network of loose connective tissue devoid of blood vessels and closely adherent to (3) the pia mater, which is the innermost thin membrane that contains blood vessels. The dura mater, arachnoid, and pia mater together form the leptomeninges.

The choroid plexuses found in the roof of the third and fourth ventricles and in part of the walls of the lateral ventricles are lined by specialized epithelial cells that differ from ependymal cells. Cuboidal cells of the choroid plexus are arranged in a single layer and many blood vessels occupy its folds. The choroid plexus is the source of cerebrospinal fluid.

Special Sense Organs
(Figures 20–31)

EAR

The ear may be divided into three parts: external ear, middle ear, and inner ear. The pinna and external auditory meatus comprise the external ear. The middle ear consists of the tympanic membrane, bony ossicles, and the auditory or eustachian tube, connecting the ear with the oropharynx. Included in the inner ear are the bony labyrinth and spiral cochlea, which contains the organ of Corti.

The outer third of the external ear is composed of elastic cartilage covered by hairy skin in which sebaceous and ceruminous glands are distributed.

Simple squamous epithelium lines the tympanic cavity of the middle ear, which contains the three bony ossicles: stapes, malleus, and incus. The tympanic membrane is an oval, semitransparent membrane that is composed of two layers of collagenous fibers and fibroblasts similar to a tendon plus epithelium on the inner and outer surfaces. It divides the middle ear from the inner ear.

Both elastic and hyaline cartilage are found in the auditory tube from its beginning to its terminus at the oropharynx. It is lined by mucous membrane with many folds. At the bony tympanic portion the epithelium is low, ciliated columnar but as the tube approaches the cartilaginous oropharynx the epithelium becomes pseudostratified with tall columnar cells, many bearing cilia.

The inner ear is divided into a bony labyrinth containing a clear fluid, perilymph, lined by endothelium, and a membranous labyrinth that contains endolymph. Within the bony portion are the vestibule, three semicircular canals filled with perilymph, and the cochlea. Enclosed in the membranous labyrinth are the utricle, utriculosaccular duct, the saccule, endolymphatic sac and duct, three membranous semicircular canals containing endolymph, and the cochlear duct (scala media). In each semicircular canal are cristae—ridges lined by an epithelium of tall columnar cells of two types, hair cells and sustentacular cells. In each utricle and saccule are maculae similar to the cristae with both hair and supporting cell types. In each macula are tiny crystals of a mixture of calcium carbonate and protein that are called otoliths. The cristae and maculae are predominantly involved with maintaining equilibrium.

ORGAN OF CORTI

On the floor of the cochlear duct is the organ of Corti, resting on a basement membrane of low cuboidal cells. This structure extends from the bony spiral lamina to the spiral ligament, and consists primarily of hair cells and supporting cells. One particular structure is the tectorial membrane, a thin gelatinous membrane in intimate contact with the cilia of hair cells. Sound waves transmitted to the organ of Corti by the endolymph of the cochlear duct from the perilymph of the scala vestibuli activate the hair cells. Nerve fibers around the bases of the hair cells receive the stimulus and transfer it to nerves in the spinal ganglion and finally, by way of bipolar cells, to the cochlear division of the acoustic nerve.

Many strains of mice are affected by mutants involving the ear structures. Malformed cristae and absence of otoliths produce conditions resulting in imbalance. Other strains are born with defects that eventually produce deafness owing to degenerating hair cells and a malformed organ of Corti. Waltzing mice, twirlers, and shaker mice fit into this category.

EYE

The principal structures of the eye consist of three layers constituting the wall of the eye: the supporting layer, middle layer, and retinal layer. The sclera or "white" of the eye is the primary supporting structure of dense connective tissue. The cornea, forming the bulging anterior portion of the eye, functions primarily as a transparent tissue transmitting light rays through the lens to the retina. However, in the rodent it also provides support, covering about half of the surface of the eyeball, and is considered a part of this layer.

The vascular middle layer or uvea provides nourishment to ocular structures and contains smooth muscle. It includes the pigmented iris and choroid, the ciliary process and ciliary body, and the lens, which is characteristically spherical in the rodent, filling almost all of the eyeball.

The retinal layer is divided into two layers; the outermost is pigmented and the inner or nervous portion contains photoreceptors, cones, and, in the mouse, a few rods.

The jellylike vitreous humor fills the interior of the eye between the lens and retina. The anterior and posterior chambers are filled with aqueous humor, secreted by the ciliary processes.

CORNEA

The cornea is a transparent avascular structure consisting of four identifiable strata. A stratified squamous epithelium covers the cornea, resting upon a basement membrane. Below is a substantia propria of specialized, dense connective tissue containing a few flattened fibroblasts dispersed among collagen fibers bound together by an amorphous cement substance. Behind this layer is Descemet's membrane, a band of refractile, homogeneous, elastic substance adjacent to a very thin squamous epithelium.

The choroid lying internal to the sclera is heavily pigmented and contains elastic fibers in addition to a single layer of capillaries.

The nervous portion of the retina consists of nine layers made up of cell bodies of many neurons and ganglia in addition to the rod and cone special photoreceptors.

The spherical lens lies just behind the pupil between the anterior and posterior chambers of the eye. This transparent structure contains a highly refractive capsule that coats the outer surface of the lens epithelium of flat cuboidal cells. Lens fibers constitute the substance of the lens, forming elongated prisms. There is no epithelium in the capsule covering the posterior portion.

Mutations resulting in a wide variety of eye defects—such as microophthalmia, cataracts, and abnormal lens formation and development—are common findings in several strains of mice. Incomplete retinal development has resulted in either abnormally short rods or absence of rods.

HARDERIAN GLAND

Partially surrounding the eyeball posteriorly is this tubuloalveolar gland. It is covered by a thin capsule whose strands divide the gland into lobules. The pyramidal epithelial cells contain a nucleus with two or more nucleoli lying along the base and lipid-containing cytoplasm, which, in prepared sections, appears vacuolated. The lamina propria of fibrous connective tissue contains melanocytes filled with pigment. A single excretory duct is present. The gland produces an oily secretion for lubricating the surface of the eye.

LACRIMAL GLAND

Paired tubuloalveolar lacrimal glands lie in two areas, extraorbital and intraorbital. The extraorbital gland is below and in front of the ear. The intraorbital gland is located where the excretory duct opens into the conjunctival sac. The pyramidal secretory cells have spherical nuclei lying near the base and intensely basophilic cytoplasm. Myoepithelial cells are present between the epithelial cells and the basement membrane. The serous secretion moistens and lubricates the surface of the eyeball and the eyelids.

Endocrine System
(Figures 32–43)

The endocrine glands include the pituitary, thyroid, parathyroid, adrenal, pineal, ovary, placenta, testis, and islets of Langerhans in the pancreas. Endocrine features of the ovary and testis will be discussed under the reproductive systems of the male and female.

PITUITARY GLAND

The pituitary gland (hypophysis) lies within the bony sella turcica of the floor of the skull, the dorsal basisphenoid bone. It consists of a pars distalis or anterior lobe, pars intermedia or intermediate lobe, and pars nervosa (neurohypophysis) or posterior lobe.

The anterior lobe is very vascular with many capillaries. Histologically, three types of cells may be identified: (1) Acidophils contain small, round nuclei and eosinophilic cytoplasm and are characterized by elaborate growth and the production of lactogenic hormones. (2) Agranular chromophobes whose cytoplasm does not stain may be stem cells. (3) Basophils may be divided into two classes: beta basophils, whose granules are aldehyde–fuchsin-positive and secrete thyrotrophic hormone, and delta basophils, whose granules are aldehyde–fuchsin-negative but stain positive with periodic acid–Schiff (red) and produce gonadotrophins, follicle-stimulating hormone (FSH), and luteinizing hormone (LH).

The intermediate lobe is separated from the anterior lobe by a cleft lined by cuboidal epithelium. Cells in this lobe have intensely staining oval nuclei and light basophilic cytoplasm. The hormone elaborated by this lobe is melanin-stimulating hormone (MSH).

The posterior lobe is filled with nonmyelinated nerve fibers, neuroglial cells (pituicytes), and connective tissue supporting capillaries. Two hormones stored and released from the pars nervosa are oxytocin and vasopressin.

THYROID GLAND

The characteristic arrangement of various-sized spherical follicles containing eosinophilic colloid identifies the thyroid gland. The follicles are lined by simple cuboidal epithelium. Increases in the height of these cells as well as changes in the staining quality and amount of colloid materials indicate levels of secretory activity. Such changes occur as an aging process in some strains of mice in which tall columnar follicle cells decrease in cell size in young mice, and increase in follicle size in mature mice. Senile changes include increased fibrous connective tissue between follicles and merging of follicles to form very large structures with flattened epithelium. The follicular cells produce the colloid substance, which contains thyroxin and triiodothyronine.

"Ultimobranchial bodies," "parafollicular cells," and C cells are synonymous terms for lightly staining cells distributed throughout the thyroid connective tissue adjacent to but outside the follicular cells. These cells are producers of thyrocalcitonin, a hormone that acts in collaboration with the parathyroid gland in maintaining calcium homeostasis in body tissues and blood. Parafollicular cells have been described in most mammals, including the mouse.

PARATHYROID GLANDS

The parathyroid gland, consisting of two or more lobes, is embedded within the thyroid gland but separated from it by a capsule of connective tissue. The cells form masses of densely packed groups interspersed with capillaries or sinusoids. At least two cell types can be identified. These chief cells consist of the dark, actively secreting cells and the light, inactive cells. A third type (similar to the oxyphil cell) has been described by some investigators.

PINEAL GLAND

The pineal gland lies on the dorsal surface of the brain at the junction of the cerebral cortex and cerebellum. A very thin capsule surrounds the gland and merges with the choroid plexus. Two types of cells can be identified: cells that originate from neuroectodermal cells and form parenchymal cells of the body of the gland, and neuroglial cells. Parenchymal cells are spherical with slightly basophilic cytoplasm. Neuroglial cells support the parenchymal cells. Mesenchymal cells in the pia mater covering give rise to connective tissue of the capsule, which provides incomplete partitions that divide the body into lobules.

ADRENAL GLAND

The adrenal glands are paired, each gland located adjacent to the anterior pole of the kidney. The adrenals are larger in the female and occasionally show strain differences. A fibrous connective tissue capsule surrounds the gland, which is divided into a cortex and medulla. In most mammals the cortical division can be subdivided into three zones of epithelial cells, but in mice only two zones are clearly seen. The subcapsular zona glomerulosa consists of small cells with large nuclei arranged in the form of arches, with capillaries between the cells. Beneath the glomerulosa is the wide zona fasciculata, which consists of columns of cells separated by connective tissue. These cells are large with vesicular nuclei and contain lipid material, appearing foamy in sections when lipid is removed in processing. The zona reticularis identified in most mammals is rarely seen in the mouse. However, a zone between the cortex and medulla is identified as the X zone in males and females before sexual maturity. The zone disappears in sexually mature males and with the first pregnancy in the female. It has been observed to persist in virgin females for varying lengths of time and in castrate males. In aging mice of certain strains (RFM, BALB/c, etc.) large foamy cells with brown pigment form in the cortical zone adjacent to the medulla. The pigment is similar to ceroid in the ovary and is referred to as "brown degeneration." In some strains of mice the change is more pronounced.

Clusters of polyhedral cells that produce adrenaline and noradrenaline form the medulla. Two cell types, the sites of origin of these hormones, have an affinity for potassium bichromate, and brown cytoplasmic granules can be seen following treatment with this chemical—the result of the chromaffin reaction.

Reproductive System
(Figures 44–76)

FEMALE REPRODUCTIVE SYSTEM

Ovary

The paired ovaries are enclosed by thin connective tissue capsules. Although in most mammalian ovaries a cortex and medulla are clearly identified, in the mouse these areas are not easily differentiated. The surface of the ovary is covered by a single layer of cuboidal cells called the "germinal" epithelium. Beneath the germinal epithelium are developing follicles and corpora lutea, interspersed with interstitial cells and blood vessels. The follicle is composed of an oocyte enclosed by a jellylike substance called the zona pellucida and multiple layers of granulosa cells. During follicular maturation the oocyte enlarges, the granulosa cells proliferate, and a fluid-filled space (antrum) appears as the follicle continues to enlarge.

At ovulation the secondary oocyte breaks through the ovarian surface and moves into the periovarian space. The remaining follicular cells (granulosa cells) undergo various changes to form a corpus luteum.

In addition to indistinct areas in the ovary, the mouse organ is unique in that corpora lutea may persist through several estrous cycles. In addition, cysts, which enlarge and become quite numerous in some strains of mice, often occur in the hilus of the vascular medulla.

Uterus

The uterus is a bicornuate tubular structure. The endometrium or mucosa of the uterus is composed of a simple columnar epithelium that lines the uterine lumen, branched tubular glands, and vascular connective tissue. The myometrium consists of an inner circular and outer longitudinal layer of smooth muscle separated by very vascular, loose connective tissue. An outer thin serosa covers the uterus. At the junction of the uterine horns with the body, a partition is present in the midline that extends through the lumen to the vagina. As the uterine lumen approaches the vaginal terminal, the epithelium changes from columnar epithelium to stratified squamous epithelium, which is continuous with the vagina.

Vagina

The vagina, a nonglandular tubular organ, consists of a stratified squamous epithelium, fibrous lamina propria, and thin muscular layer. During the estrous cycle the vaginal epithelium also undergoes cyclic changes, characterized by proliferation of the basal cells, cornification, leukocytic infiltration, and variation in the number of epithelial cell layers.

As the phases of estrus progress, histological examination may indicate overlapping of one phase with another. The predominating phase may be determined by the abundance or scarcity of cornified cells or mitotic figures, the presence or absence of leukocytes, and the thickness of the epithelial cell layers. Estrus consists of five phases, two of which—proestrus and estrus—represent growth, and two—metestrus I and II—in which degenerative changes in the epithelium occur. Diestrus represents a transition between degeneration and renewed growth, i.e., rebuilding the epithelial lining. Proestrus is characterized by an increase in cell layers, cornifying granulosa layers, active cell division, and a scarcity of leukocytes. As the estrus phase predominates, the cornified layer becomes superficial prior to desquamation, the number of mitoses are reduced, and leukocytes disappear. Metestrus I is identified by desquamation of cornified cells into the vaginal lumen and the appearance of leukocytes in the outer layer of the epithelium. Metestrus II includes a marked increase in leukocyte numbers, reduction in epithelial cell layers, and the presence of an occasional mitotic figure. The onset of diestrus marks a return to active growth of the now thin epithelium and is characterized by an increase in numbers of cell layers, reduced numbers of leukocytes, and the appearance of a few dividing cells. Large, mucoid, PAS-positive cells may appear between the proestrus and estrus phases, probably stimulated by progesterone. The mucous cells rest superficially on the outer cornified layer, desquamating just prior to that layer.

Placenta

The placenta is a flat, spherical structure that is smooth on the fetal side where it is covered by cuboidal amitotic epithelium. This rests on the chorion, a layer of embryonic connective tissue. Most of the placenta is composed of chorionic villi bathed in a pool of maternal blood. The trophoblast covers the villi, the cores of which contain fetal blood vessels. The trophoblast is divided into a syncytial and deeper cellular layers. Giant cells are easily identified in the syncytiotrophoblast of the mature placenta. Thick portions of this region are the sources of placental hormones: estrogen, progesterone, lactogen, and chorionic gonadotrophin. The maternal portion of the placenta is the decidua basalis. The decidual cells are large polygonal cells and are rich in glycogen and lipids. Irregular masses of an eosinophilic, homogeneous substance called fibrinoid are found in young animals and increase with the age of placenta. Endometrial glands, fibrinoid, and some trophoblast cells are also present.

Mammary Glands

Mammary tissue develops in both the male and female embryos. In the male, because of the inhibitory effect of testosterone secreted by the fetal testis, the mammary tissue does not mature into nipples and ducts.

In the female embryo, mammary buds appear by the fifteenth day within the epidermis and gradually lengthen into epithelial cords. A few days later, tubules form within these cords. The epidermal cells surrounding the cords develop into nipples. At birth the female gland consists of a nipple opening into a primary duct. This duct branches into secondary ducts, which lie within adipose tissue and are lined by low cuboidal epithelium. A few myoepithelial cells lie between the epithelium and basement membrane. Increasing ductal branching occurs by the fourth week as the glands rapidly grow. By 6 months alveoli develop, although the degree of development varies with mouse strain.

During pregnancy and lactation the number of alveoli increases, with hypertrophy of epithelial cells accompanying widening of the lumen. As milk is secreted in the alveoli, the cuboidal cells are flattened into a squamous pattern. At cessation of suckling, alveolar epithelial cells resume cuboidal shape.

MALE REPRODUCTIVE SYSTEM

Testes

A thick capsule, the tunica albuginea, covers each testis. The internal structure consists of tubules separated by thin septa. Inside are the highly convoluted seminiferous tubules, which are separated by loose connective tissue with some elastic fibers and the cells of Leydig. The tubules are lined by a highly specialized stratified epithelium, seminiferous or germinal epithelium, which rests on a basement membrane. The cells touching on the basement membrane are of two types, spermatogonia and Sertoli or sustentacular cells. The Sertoli cells are large pale cells with vesicular nuclei and a prominent nucleolus. These cells lie against the basement membrane and are believed to nourish the germ cells.

Spermatogenic cells occur in layers. The first and most primitive are types A and B spermatogonia, which are dispersed among the Sertoli cells lying against the basement membrane. Upon mitotic division the spermatogonia form primary spermatocytes, which undergo meiosis to form secondary spermatocytes. These cells, after dividing mitotically, form spermatids, small cells with eccentric nuclei and a tail, many of which are in direct contact with Sertoli (nurse) cells while transforming into immature spermatozoa. Each mature sperm contains a sickle-shaped head (though width of the head varies considerably in several strains of mice), midpiece, and tail. These cells now lie next to the lumen.

The interstitial tissue of the testis is the endocrine portion, lying between the seminiferous tubules. The loose connective tissue contains blood and lymphatic vessels, nerves, and groups of epithelial cells, the cells of Leydig, which are the source of testosterone. The oval cells contain spherical to oval nuclei and cytoplasm, which is often vacuolated. Removal of lipid material during preparation of the tissue results in the formation of cytoplasmic vacuoles.

Epididymis

The epididymis is covered by a capsule of fibrous connective tissue. It is divided into a head, body, and tail. Most of the efferent ductules are located in the head. The greatly coiled duct of the epididymis occupies the body and tail. The efferent ductules are lined by low to tall columnar epithelial cells with stereocilia projecting into the lumen.

Vesicular Glands (Seminal Vesicles)

These elongated and curved paired glands lie immediately above the bladder. Internally the alveolar mucosa in multiple folds is lined by tall columnar epithelium whose cells shorten as the gland becomes distended with acidophilic secretion.

Vas Deferens

This thick-walled tube is lined by pseudostratified columnar epithelium. Loose connective tissue with many elastic fibers surrounds the star-shaped lumen. The muscularis is thick with smooth muscle, composed of inner longitudinal, middle circular, and outer longitudinal layers.

Coagulating and Ampullary Glands

Immediately below and attached to the vesicular glands are the paired coagulating glands whose folded mucosa is lined by columnar cells with centrally located spherical nuclei and acidophilic cytoplasm.

Lateral to the vesicular glands is a group of small glands with many short ducts forming the ampullary glands. Their epithelium is low columnar with large oval nuclei, and their secretion is eosinophilic.

Dorsal and Ventral Prostate

Immediately below the bladder are the dorsal and ventral prostate glands. Both glands have several ducts that enter the urethra. Ducts from the dorsal gland enter laterally; the ventral ducts enter ventrally. The glands are surrounded by smooth muscle and a fibromuscular stroma. The secretion often condenses, forming globules that give a granular, acidophilic appearance.

Bulbourethral Glands (Cowper's Glands)

These paired glands lie approximately at the junction of the membranous urethra and penis on the lateral sides. Each gland is divided into a body and a tail, both of which are embedded in skeletal muscle. The glands consist of tubular alveoli, which are lined by tall columnar cells of variable height. The small, dark nuclei lie flattened in the cell with basophilic cytoplasm and rest on a well-developed basement membrane.

Preputial Glands

The flat broad cells of this gland are identical in appearance and function to sebaceous glands. Characteristically, the nucleus degenerates and becomes part of the oily secretion, which is then excreted. The duct of the gland is lined with stratified squamous epithelium; it opens into the preputial space. Occasionally, pigment is observed in the surrounding connective tissue.

Digestive System
(Figures 77–104)

TONGUE

The tongue consists principally of striated muscle covered by a mucous membrane. Muscle fibers run longitudinally and horizontally. In the fibroelastic tissue between muscle bundles, blood vessels and nerves are found. The epithelium of the mucosa is thick, stratified squamous and is cornified on the dorsal surface.

On the rough dorsal surface of the tongue are areas of various papillae projecting from the surface. Four types may be identified: foliate, vallate, filiform, and fungiform. Filiform (cone-shaped) papillae have a broad base and pointed tips, which project posteriorly. A few fungiform papillae, somewhat square in shape with a connective tissue core, are among the filiform papillae. Several rows of foliate papillae are found along the epithelial surface of the sides of the tongue opposite molar teeth. Behind the molars, a single, roughly square vallate papilla occurs, bounded by deep furrows on either side in the tongue surface.

Taste buds are located on both sides of the dorsal surface of the vallate papilla and on the surface of the foliate papilla. Oval taste buds are arranged perpendicular to the surface of the mucous membrane epithelium of the mouth. Taste buds contain two types of cells, sustentacular or gustatory cells and neuroepithelial cells. Sustentacular cells are elongate and somewhat banana-shaped and follow a curved course in the taste bud. They are arranged so that they are pointed toward a small opening in the bud, the pore, which empties into the lumen of the papilla, which in turn opens on the dorsal surface of the tongue. In the central part of the taste bud, sustentacular cells mix with neuroepithelial cells. These cells are long and narrow, with darkly staining nuclei that end in tiny hairs projecting from the pore.

SALIVARY GLANDS

There are three salivary glands in the mouse: parotid, submandibular, and sublingual. They are compound tubuloalveolar glands divided into lobules by connective tissue septa and surrounded by a capsule. Tall pyramidal cells line the secretory alveoli and lie on a basement membrane. Myoepithelial or basket cells lie between glandular cells and the basement membrane.

The parotid gland produces a serous fluid. The cells have a spherical, centrally located nucleus with basophilic cytoplasm. The short, narrow ducts extending from the gland are lined by low cuboidal epithelium.

The secretory alveoli of the submandibular gland are lined by tall pyramidal cells that produce mucus and serous fluid. In male mice the intercalated tubules and some alveoli have tall columnar cells whose nuclei lie at the base of the cell. Abundant eosinophilic granules fill the cytoplasm. Nuclei in cells of the female are centrally located and only a few cytoplasmic granules are seen. They increase in number as young females as well as young males mature.

In the sublingual gland, tall pyramidal cells whose nuclei lie basally have pale-staining cytoplasm. Compared with the other two glands these mucous-secreting cells appear pale, bluish white in hematoxylin-stained preparations. The short intercalated ducts are lined by low cuboidal epithelium. Stratified squamous epithelium lines the chief excretory duct.

ESOPHAGUS

The esophagus has a folded mucosa lined by cornified, stratified squamous epithelium. No glands are present submucosally. The muscle layer consists of striated muscle throughout the length of the esophagus.

STOMACH

The stomach is divided into two approximately equal nonglandular and glandular portions by a ridge. Cornified, stratified squamous epithelium lines the nonglandular stomach. Mucus-secreting cells, zymogenic or chief cells, and acid-secreting parietal cells comprise the glandular portion.

SMALL INTESTINE

Three regions of the small intestine may be identified: duodenum, jejunum, and ileum. All three have characteristic villi, fingerlike projections that increase the surface area. Columnar epithelial cells cover the villi. Their nuclei are located at the base, while a striated border is seen along the lumen. Goblet cells filled with mucus are distributed among the epithelial cells, increasing in number in the ileum. The villi decrease in size from the duodenum to the ileum. Between adjacent villi are the crypts of Lieberkühn. Glands of Brunner are present only in the duodenum, extending below the bases of the villi into the submucosa. Paneth cells with prominent eosinophilic granules are found at the bases of the crypts, especially in the jejunum. Lymphoid nodules (Peyer's patches) are distributed in the submucosa throughout the small intestine.

LARGE INTESTINE

The large intestine is characterized by columnar epithelium with many goblet cells and a lack of villi. In the anal region the crypts of Lieberkühn disappear and a transitional zone between columnar and stratified squamous epithelium occurs between mucous membrane and skin. The opening is surrounded by the external sphincter (striated) muscle and sebaceous glands.

ENTEROCHROMAFFIN SYSTEM

Cells of the enterochromaffin system have been described throughout the gastrointestinal system. They are characterized by the presence of cytoplasmic granules that react to treatment with silver salts and with bichromate. These cells are known to contain serotonin (5-hydroxytryptamine).

LIVER

The liver is divided into indistinct lobules having in their center hepatic (central) veins from which cords of hepatic cells, separated by sinusoids, radiate out to the periphery. The outer boundaries of the lobules are poorly demarcated by the portal triads (hepatic artery, portal vein, bile duct). Reticular fibers, originating from the fibrous capsule (Glisson's capsule) covering the liver, penetrate the organ supporting the parenchyma.

Interlobular branches of the hepatic artery and portal vein provide the afferent blood supply to the liver sinusoids. The sinusoids in turn drain into the central veins of the lobule, which are branches of the hepatic vein. The sinusoids are lined by endothelium and specialized reticuloendothelial cells, the Kupffer cells, that function as phagocytes.

Between adjacent hepatocytes, on the opposite side of the sinusoid, are bile canaliculi. These extend from the central vein outward to the periphery of the lobule, where they open into bile ducts lined by cuboidal epithelium. These bile ducts carry bile to the gallbladder.

The hepatic cells or hepatocytes are large polygonal cells with a prominent central nucleus and eosinophilic cytoplasm. Occasionally two nuclei and one or more nucleoli may be seen.

GALLBLADDER

The gallbladder, a sac for bile storage, consists of a folded mucosa lined by simple columnar epithelium, beneath which lies the vascular lamina propria. This is surrounded by smooth muscle and an outer serosa. The common bile duct leading from the gallbladder to the intestine is lined by cuboidal epithelium, which changes to columnar as it approaches the duodenum.

PANCREAS

The pancreas is both an exocrine and an endocrine gland. The exocrine pancreas consists of cells that form tightly packed acini. These acinar cells are pyramidal in shape, with their apical ends bordering the lumen and broad, intensively basophilic bases resting on a basement membrane. The apical cytoplasm of these cells contains eosinophilic zymogen granules.

The endocrine pancreas consists of clusters of endocrine cells with their capillary network, the islets of Langerhans, which are embedded in acinar tissue. Alpha cells that secrete glucagon occupy the periphery of the islet while beta cells that secrete insulin are more numerous and fill the central area of the islet.

Urinary System
(Figures 105–111)

KIDNEY

The paired, bean-shaped kidneys are covered by a capsule of connective tissue. The shorter concave border is called the hilus; from it extend the renal artery and vein, nerves, and the ureter on its way to the bladder.

The kidney is divided into two regions, an outer cortex containing tubules and glomeruli and an inner medulla composed primarily of blood vessels and collecting tubules. The functional unit of this organ is the nephron, consisting of a glomerulus and associated urinary tubules lined with cuboidal epithelium. The glomerulus, composed of a tuft of capillaries surrounded by Bowman's capsule, is supplied by an afferent arteriole and drained by an efferent arteriole. In the mouse a marked sex difference is noted in the type of epithelium of Bowman's capsule: cuboidal epithelium for most glomeruli of the male, squamous epithelium in the female. The segment of nephron that emerges from a glomerulus and courses through the cortex is the proximal convoluted tubule, lined by cuboidal epithelium with a pronounced brush border. The tubule continues into the medulla as the descending portion of the loop of Henle and returns to the cortex as the ascending portion of this loop. From this structure the distal convoluted tubule empties into the collecting tubules located within the medulla. Although cuboidal epithelium lines the distal tubules, the brush border is absent. The collecting tubules with columnar epithelial cells connect with the ureter and the epithelium becomes transitional (urothelium).

BLADDER

The pear-shaped bladder is lined with transitional epithelium (urothelium) of two to four cell layers, depending upon the degree of distention. The lamina propria of the bladder is very vascular and is surrounded by a thick muscularis composed of several layers of smooth muscle. This muscle forms a sphincter at the bladder neck, where it joins the membranous portion of the urethra.

URETHRA AND PENIS

The membranous urethra extends from the neck of the bladder to the pelvic girdle, where it enters the penis. It is lined by transitional epithelium, and its vascular lamina propria forms a sheath around the urethra. Embedded in this structure are the urethral glands of Littré, clusters of secretory alveolar cells with oval nuclei and basophilic cytoplasmic secretion. In the penis, the corpus cavernosus urethrae is formed by a sheath of smooth muscle and an outer fibrous connective tissue sheath. Cavernous spaces lined by endothelium form from tissues of the inner and outer sheaths. Blood fills these spaces, producing erection of the penis. Most of the urethra from the bladder to the tip of the penis is lined by transitional epithelium (urothelium), which changes abruptly at the orifice into stratified squamous. Dorsally, in the septum between the paired corpora cavernosa, is a small compact bone, the os penis.

Skin
(Figures 112–122)

The skin consists of two distinct components: the epidermis of keratinizing, stratified squamous epithelium, and the dermis. Thin skin occurs over most of the hairy regions while thick skin appears over the hairless areas, i.e., feet, tail, nipples, and genital and anal areas.

Thick skin consists of four layers, the deepest consisting of a single layer of cells with oval nuclei and indistinct cell membranes resting on a basement membrane immediately adjacent to the dermis. This is the stratum germinativum, in which mitotic figures are often seen and whose cells replace cells of the outer layers. The stratum spinosum, four to five cells thick, is the next layer. These cells are connected across intercellular spaces by tonofibrils. The stratum granulosum is identified by cells containing keratohyalin granules easily stained with hematoxylin. The outermost layer of desquamating dead cells is the stratum corneum.

The dermis consists primarily of connective tissue as well as many blood vessels, nerves, collagen and elastic fibers, and fat cells. Thin strands of smooth muscle (arrector pili) associated with hair follicles and striated muscle (panniculus carnosus) are also found in the dermis. Pigmented areas contain melanocytes. Dermal papillae containing blood vessels and nerves project up into the epidermis. Hair projects from the skin surface; the follicle containing the shaft, bulb, and matrix is located in the dermis. Actively dividing epidermal cells in the bulb produce growth of the shaft. Sebaceous glands accompany each hair follicle, as does an arrector pili. Pigmented hair results from melanocytes in basal epidermal cells.

Specialized tactile hairs are found abundantly distributed about the lips, cheeks, and mouth. These elongated whiskers are called vibrissae. Although their structure is similar to that of hair follicles, the vibrissae are much larger and are surrounded by venous blood sinuses and nerve fibers. Striated muscle fibers replace smooth muscle, and a sebaceous gland accompanies each vibrissa.

Sebaceous glands have ducts that open along the hair shaft. The gland cells are large, oval cells with prominent nuclei in the basal layer. As the secretion accumulates in the gland, the cells lose their nuclei and become part of the secretion, which is called sebum.

Respiratory System
(Figures 123–132)

The respiratory system includes the nasal passages, pharynx, larynx, trachea, and lung. The epithelium of the nasal passages is of the pseudostratified, ciliated columnar type with numerous goblet cells. The nasal passages also contain olfactory epithelium, including specialized bipolar nerve cells supported by two cell types, basal and supporting cells. The supporting cells form tall columns of cells with oval nuclei near the surface; many olfactory nerve cells are subjacent. The olfactory cells are bipolar ganglion cells, the apical portion forming a dendrite that extends to the surface as a base for nonmotile cilia. Below the nucleus the cytoplasm extends to a fine thread to form an unmyelinated axon that eventually constitutes, along with others, the olfactory nerve bundles. Basal cells are roughly triangular and are seen between other cells along the basement membrane. Glands in the interstitial tissue membrane are the tubuloalveolar type (mucous and serous).

The pseudostratified, ciliated columnar epithelium and glands of the nasal mucosa continue into the nasopharynx, interrupted by stratified squamous epithelium as the oropharynx and larynx are crossed, and continue down the major portions of the respiratory tract.

The trachea is lined by a mucous membrane with pseudostratified, ciliated columnar epithelium with numerous mixed glands and goblet cells. The tube is prevented from collapsing by the presence of ten to twelve C-shaped hyaline cartilage rings. The trachea bifurcates into bronchi, which branch successively into bronchioles. In the mouse there are no bronchial glands, nor is there cartilage beyond the bifurcation of the trachea.

In the lungs the bronchioles are lined by epithelium consisting of ciliated and nonciliated cells, the latter known as Clara cells. Clara cells are identified in thin methacrylate sections by characteristic bulging into the lumen and absence of cilia. The bronchioles branch successively to form alveolar ducts, alveolar sacs, and finally alveoli.

The alveoli are thin-walled tubes lined by epithelial cells and supported by interstitial tissue of the alveolar wall (septa), which consists of a few reticular fibers, septal cells, and many capillaries. The alveolar epithelial cells are of two types, the squamous or type I epithelial cell and the great alveolar or type II epithelial cell. The type II cell can be recognized in thin methacrylate sections by its location at the junction of alveolar septa, its cuboidal shape, and its vacuolated cytoplasm (loss of lamellar bodies during tissue preparation).

Lymphatic System
(Figures 133–140)

The lymphatic system includes lymph nodes and nodules present in areas of the digestive tract as well as lymph and lymphatic vessels that carry the fluid from tissue spaces to the circulatory system. Tonsils are absent in the mouse.

Lymphatic vessels are lined by a single layer of endothelial cells. Collagen and elastic fibers as well as smooth muscle fibers envelop larger vessels. In the hilus of lymph nodes one can identify cross sections of efferent vessels, while afferent lymphatics are found entering the node at the peripheral marginal sinuses.

Lymph nodes are surrounded by a fibrous capsule and are divided into a marginal subcapsular sinus, the cortex containing dense masses of lymphocytes and the medulla composed of lymphoid elements with large sinusoids lined by littoral cells. Supporting tissue in the node consists of reticular cells and fibers. Germinal centers containing precursors of mature lymphocytes are occasionally seen in cortical lymphatic centers.

THYMUS

The thymus, a lymphoid organ, is composed of a cortex packed with small lymphocytes, giving a very dark blue appearance when stained with hematoxylin, and a lighter-stained medulla with far fewer cells and containing aggregations of a few large epithelial cells forming Hassall's corpuscle. In other animals keratin is often observed within the corpuscle but is absent in the mouse. The medulla is well supplied by blood vessels.

SPLEEN

The spleen is a very vascular organ covered with a capsule of dense connective tissue from which trabeculae project into the spleen pulp. Branches of the splenic artery and vein, nerves, smooth muscle fibers, and reticular fibers accompany the trabeculae to form supporting connective tissue for red and white pulp White pulp is lymphoid tissue and forms characteristic nodules, also called Malpighian corpuscles. Occasionally, germinal centers are seen, surrounded by a spherical mass of lymphocytes. A central arteriole occupies the center of the nodule.

Red pulp consists primarily of erythropoietic cells, lymphocytes, granulocytes. plasma cells, mast cells, and megakaryocytes. In the mouse, extramedullary hemopoiesis is commonly found in red pulp. Apparently, there are no true sinuses in mouse spleen. Some strain differences in patterns of reticulum have been described.

Bone, Cartilage, Fat
(Figures 141–149)

BONE

There are two kinds of ossification, endochondral and membranous, implying the presence or absence of a prior cartilaginous model. In both processes bone is formed by osteoblasts, which originate from mesenchyme. Osteoblasts are spindle-shaped to columnar cells with an eccentric nucleus and basophilic cytoplasm. They congregate on surfaces and lay down the organic matrix (osteoid), which then becomes mineralized. When the osteoblasts become trapped in small spaces (lacunae) within this osteoid, they are called osteocytes.

Bone is resorbed by two different mechanisms: osteoclasia and osteolysis. Osteoclasia is the surface resorption of bony tissue by multinucleated, giant cells (osteoclasts) typically located in Howship's lacunae during the active phase. Osteolysis refers to a deep-seated resorption centered around the activity of old osteocytes.

Long bones and vertebrae grow from epiphyseal plates and articular cartilage. In the growing animal there are three zones in the epiphyseal plate: zone of resting cartilage (proliferation), zone of columnar cartilage (maturation), and zone of vesicular cartilage (hypertrophy). Cessation of longitudinal growth coincides with sexual maturity, and the plate gradually decreases in size. In the growing individual, articular cartilage grows in two directions. Growth toward the epiphysis mimics the growth of the epiphyseal plate, while growth toward the surface aims at replacement.

CARTILAGE

Cartilage is a supporting connective tissue consisting of three types: hyaline cartilage, elastic cartilage, and fibrocartilage. Hyaline cartilage has a white and glossy appearance in the living state and is composed of cells (chondrocytes) enclosed in lacunae. The cells are found individually or in groups within a hyaline matrix of intercellular substance. It is from this type of cartilage that some areas of bone formation are derived.

Elastic cartilage, as the name implies, contains abundant elastic and some collagen fibers within intercellular substance. Examples are the pinna of the ear, nasal cartilage, and epiglottis.

Fibrocartilage is a tough substance containing an excess amount of collagen fibers; its lacunae are arranged in symmetrical rows. It is found closely associated with tendons attached to bone.

FAT

Adipose tissue in the mouse is divided into two types: white and brown fat. Most white fat is abundantly found in several specific areas, i.e., in the mesenteries around blood vessels, kidneys, adrenals, ovaries, and testes and in the axillary and inguinal regions. In H & E-stained sections a white fat cell is identified as a round or polyhedral cell whose fat has been removed during processing and whose nucleus lies flat against the cell membrane. Some reticular fibers surround and support the capillaries among the fat cells.

Brown fat consists of groups of polygonal cells with lipid droplets in granular cytoplasm. The nucleus is centrally located. Numerous capillaries and reticular fibers support the fat cells. Brown fat is also found in specific locations, such as in lobes between scapular bones, adjacent to the thymus, and at the hilus of the kidney. The fat lobes between scapulae are called hibernating glands, although there is no evidence that these structures secrete hormones.

Muscular System
(Figures 150–153)

Two characteristics of muscle tissue are contractility and, to a lesser extent, conductivity. The muscle fibers are the units of contraction. Abundant connective tissue is found among muscle fibers as a supporting substance and also as a medium in which blood vessels and nerves are placed in intimate contact with muscle fibers.

Morphologically, muscle tissue may be classified into three types: smooth involuntary, striated voluntary (skeletal), and striated involuntary (cardiac).

SMOOTH MUSCLE

Smooth muscle is found in the walls of most tubular structures and organs (gastrointestinal tract, arteries and veins, most hollow viscera, uterus, bladder, etc.). Smooth muscle fibers are tapered and have a wide center around the eccentrically placed nucleus and pointed or blunt ends; the cells are roughly cigar-shaped. The muscle fibers are arranged in sheets or bundles surrounded by connective tissue. In a longitudinal view the nuclei appear to lie on top of the sheet of fibers. Blood vessels and nerves accompany the muscle fibers in their connective tissue covering. Generally all muscle fibers in a specific area contract as a unit, as in the intestinal tract during peristalsis.

STRIATED MUSCLE

Striated (voluntary or skeletal) muscles are capable of powerful contraction or maintaining tonus (e.g., holding the head erect). Connective tissue also accompanies this type of muscle, bearing blood vessels and nerves as in smooth muscle. However, skeletal muscle cells are larger and multinucleate (even syncytial), ranging in length from a few millimeters to a few centimeters. The nuclei are elongated and oval and appear on the outside of the muscle fiber bundle in longitudinal section.

Each fiber is enclosed by a very thin covering, the sarcolemma, which adheres very closely to the sides of the fiber and is not easily recognized in stained preparations. When viewed in longitudinal section the striations, for which this type of muscle is named, are striking. Vertical light (I) bands and dark (A) bands alternate with each other. A dark (Z) line bisects each light (I) band.

All skeletal muscle fibers have an associate nerve fiber whose cell body is located in the anterior horn of the spinal cord. Its axon extends to a terminal point on the muscle, the motor end plate. Some of these structures in the photomicrograph section have been demonstrated by silver staining.

CARDIAC MUSCLE

Cardiac (involuntary) striated muscle appears as a syncytium of fibers branching and anastomosing; between the fiber bundles is connective tissue bearing blood and lymph capillaries and nerves. Nuclei of cardiac muscle are ovoid with blunt ends and appear to be on top and in the middle of the muscle fibers. Cross striations are seen in the cytoplasm, though not stained as intensely as those in skeletal muscle. In addition to cross striations, transverse dark lines are often identified at various intervals in a steplike pattern. These are intercalated disks, and they mark the boundary between adjacent muscle fibers.

Circulatory System
(Figures 154–168)

HEART

The heart is a hollow muscular organ enclosed in a fibroelastic pericardial sac. In the wall of the heart, three distinct layers can be identified. The inner layer lining the chambers and covering the valves is the endocardium. The thick cardiac muscle of the heart is the myocardium. The epicardium is the thin layer of mesothelium and connective tissue that covers the surface of the heart as the visceral pericardium. The pericardial cavity, filled with fluid, separates the pericardium and epicardium.

The impulse-conducting system of the heart consists of specialized cardiac muscle fibers in large amounts of fibroelastic tissue abundantly supplied by nerves from the autonomic nervous system. Among the constituents of the system are the sinoatrial (SA) node (the pacemaker of the heart), the atrioventricular (AV) node, and Purkinje fibers. The latter are very wide fibers having two nuclei in a clear mass of sarcoplasm rich in mitochondria and glycogen, with cross striations but no intercalated disks. The myofibrils are peripherally located, leaving a wide central space, seen in prepared sections. The Purkinje fibers conduct impulses for ventricular contraction, reacting much more rapidly than ordinary cardiac muscle.

ARTERIES AND VEINS

The walls of arteries consist of three layers: tunica intima, tunica media, and tunica adventia. The intima consists of an endothelial lining of the lumen with subendothelial connective tissue and abundant elastic fibers. A thick wall of smooth muscle and elastic fibers constitutes the media, which varies in thickness with the caliber of the artery. The outermost layer or adventia contains connective tissue and collagen, which gradually merges into the surrounding tissues. Valves are absent in arteries.

Although similar to arteries, veins are generally thinner and have fewer elastic fibers in the intima and less smooth muscle in the media, although the adventia is very similar to that of arteries. In mice the smooth muscle layer of the pulmonary vein is replaced by cardiac muscle. Paired valves are present throughout the venous system.

CAPILLARIES

The capillary wall is composed of a single layer of endothelial cells on an underlying basement membrane with no smooth muscle but many reticular fibers.

PERIPHERAL BLOOD CELLS—WRIGHT'S STAIN

Blood smears made from tail vein preparations include monocytes, small and large lymphocytes of the agranulocytic series, and neutrophils and eosinophils of the granulocytic series. Basophils are very rarely seen. Eosinophils vary considerably in nuclear morphology, size of granules, and intensity of cytoplasmic staining. In Figures 165–168 several of these variants can be compared.

Erythrocytes appear in the background of Figures 161–168 as gray, biconcave disks of uniform size.

Platelets appear in clusters with red granules and blue cytoplasm.

Color Figures

1. Head: Sagittal section H & E 10×

The upper half of this midsagittal section through the head extends from the lips and incisor tooth on the right through the nasal cavity and its sinuses to the trachea, immediately above which is the brain. The homogeneous pink cerebral cortex contrasts with the arbor vitae or "tree of life" appearance of the dark blue cerebellum at the posterior brain, at left. The small triangular structure immediately below the posterior cerebral cortex is the pituitary gland. The pink triangle includes the intermediate lobe and neurohypophysis, which lie on top of the horizontal dark, anterior lobe. The lower jaw includes a tooth and overhanging tongue.

1. Skull
2. Skin
3. Cerebellum
4. Cerebral cortex
5. Nasal sinuses
6. Trachea
7. Mouth
8. Tooth
9. Tongue
10. Esophagus
11. Pituitary gland

2. Cerebellum and medulla Masson stain 25×

In this cross section of brain the patches of green are the granular layer of the cerebellum. Immediately adjacent are areas of reddish brown, the molecular layer on the edge of which consists of cell bodies of Purkinje cells not seen at this low magnification. The lower half of the section is occupied by the medulla, whose nerve cell bodies are located in the dark regions, while the nerve fibers course through the alternate pink areas.

1. Cerebellum
2. Nerve fibers and neuroglia
3. Medulla
4. Choroid plexus in ventricle

3. Cerebral cortex Luxol Fast Blue Nuclear Fast Red 500×

The blue-stained fibers crossing the field are myelinated fibers. Densely red-stained nuclei of neuroglial cells appear between fibers. The group of nucleated nerve cells at lower left forms a ganglion, many of whose cells have bright red nucleoli.

1. Myelinated (blue) fibers
2. Ganglion cells
3. Neuroglial cell nuclei

4. Choroid plexus H & E 250×

The choroid plexus of capillaries lies in the ventricle, lined by a specialized epithelium of ependymal cells. Blood in a venous sinusoid appears at center left.

1. Choroid plexus
2. Ependymal cell layer
3. Ventricle
4. Cerebral cortex
5. Erythrocytes in venous sinusoid

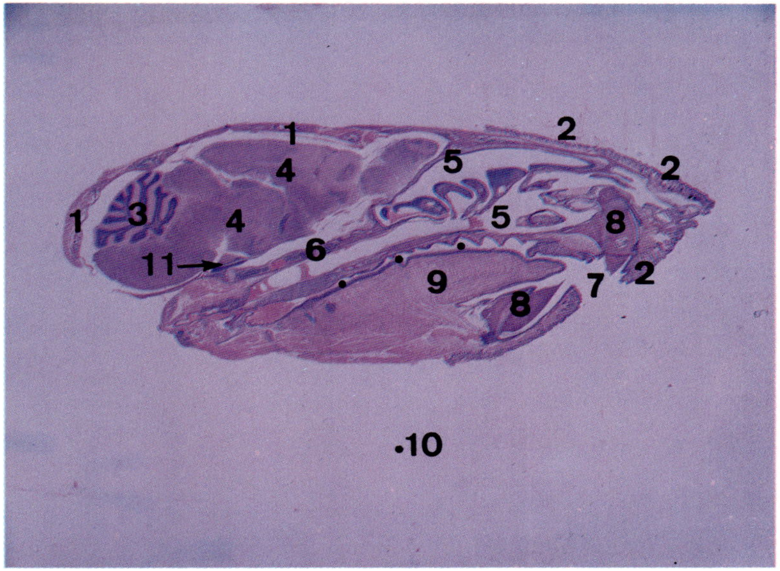

Figure 1

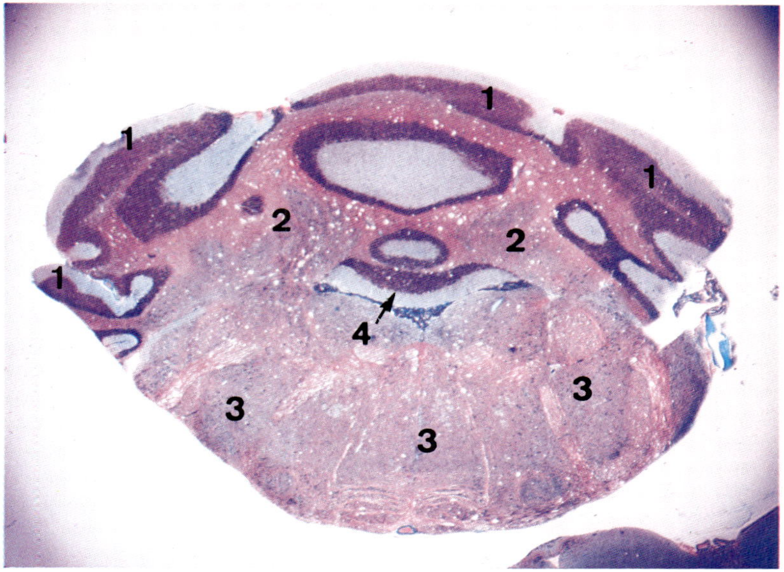

Figure 2

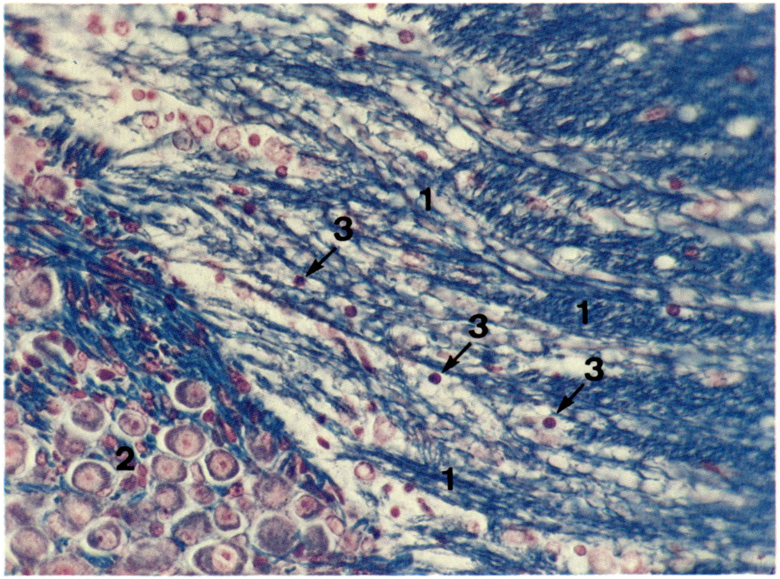

Figure 3

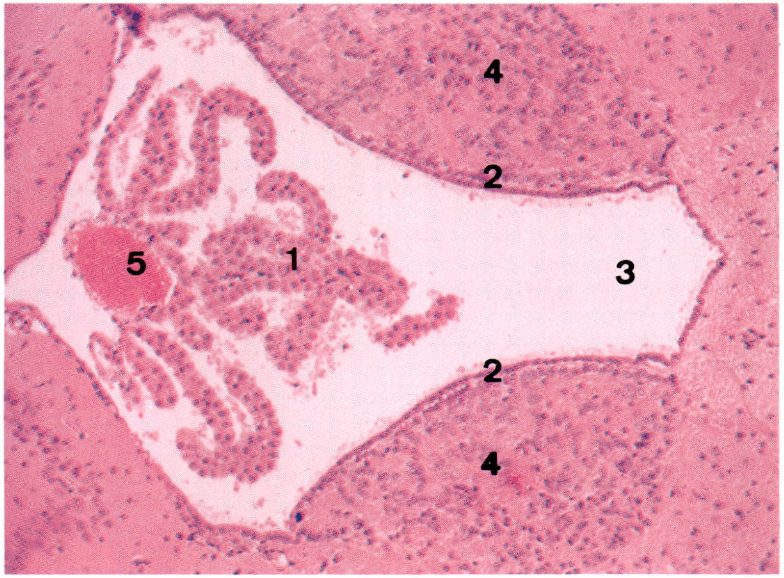

Figure 4

5. Spinal cord: Anterior horn Bodian silver 500×

The entire field is filled with various nerve fibers. Several large motor cells, two of which exhibit pronounced axons, are seen at lower left and upper right.

1. Motor cell
2. Neuroglial cell nucleus
3. Nerve fibers
4. Axon of nerve cell
5. Neurofibrillae (fine brown lines)

6. Ganglion Masson stain 500×

Characteristically large round ganglion cell bodies fill most of the field, with a vein at lower right and green bone at lower left. Two nerve trunks in cross section are present at upper right with a few Schwann cell nuclei in focus.

1. Ganglion cell bodies
2. Nerve trunk
3. Vein
4. Bone
5. Fat cells

7. Cerebral cortex Bodian silver 500×

Several large motor neurons exhibit axons, and many neurofibrillae can be seen within their cytoplasm. In the background are nuclei of neuroglial cells and nerve fibers.

1. Motor cell nucleus and nucleolus
2. Dendrite
3. Nerve fiber
4. Neuroglial cell nucleus

8. Motor neurons in cerebral cortex Bodian silver 500×

The axons and dendrites of each cell are easily identified. Within the cytoplasm are fine, dark lines, the neurofibrillae. Neuroglial cells and nerve fibers completely fill the background.

1. Nucleus and nucleolus of neuron
2. Axon
3. Dendrite
4. Neurofibrillary meshwork
5. Nucleus of neuroglial cell
6. Nerve fiber

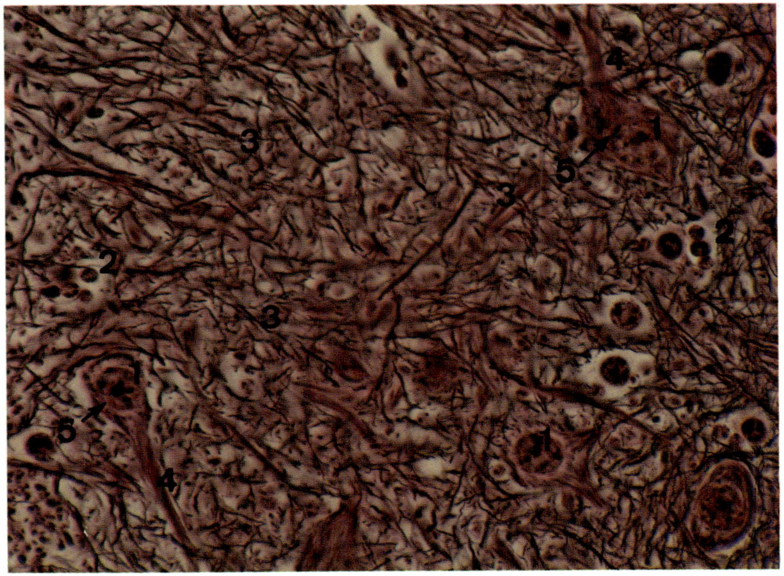

Figure 5

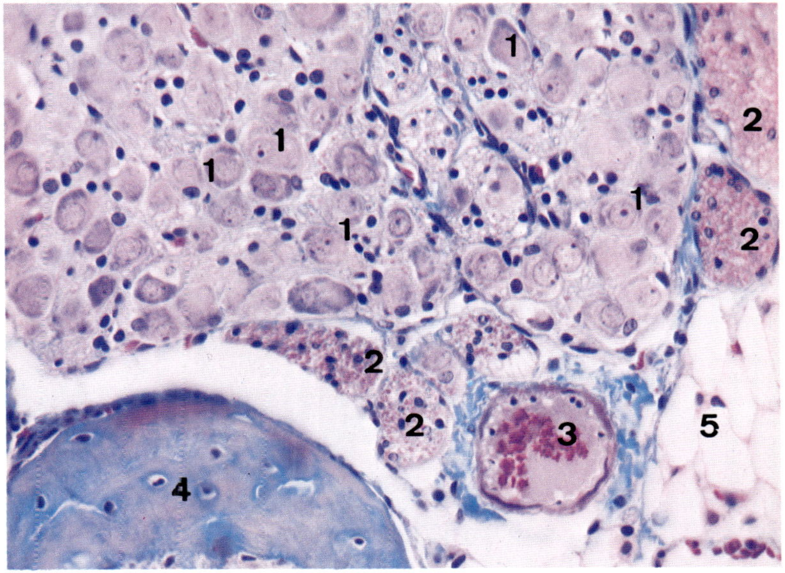

Figure 6

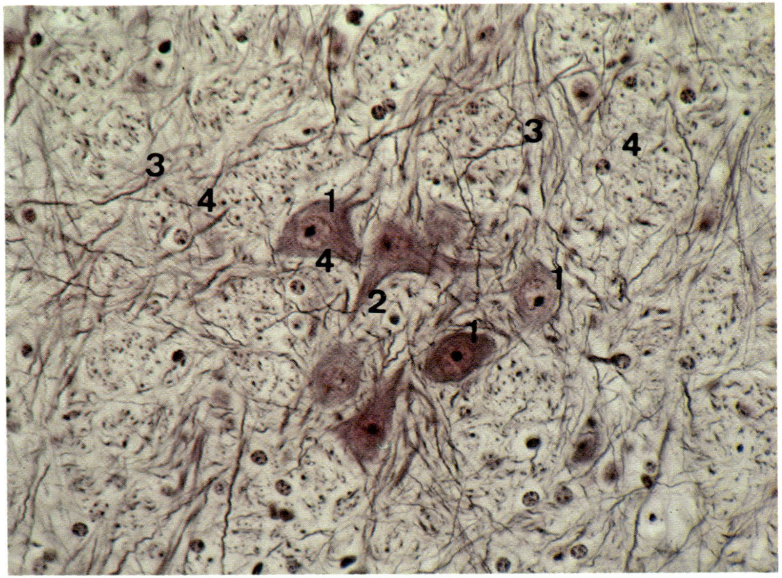

Figure 7

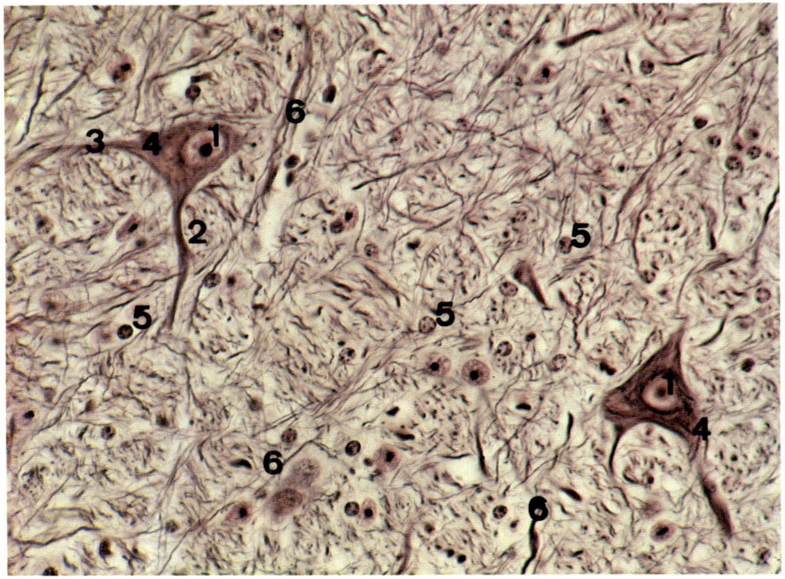

Figure 8

9. Cerebral cortex — Luxol Fast Blue, Cresylecht violet — 500×

A large motor cell body in the center foreground exhibits a nucleus with nucleolus and abundant Nissl substance in the cytoplasm. Purple myelinated nerve fibers extend across the field, among which are several darkly stained nuclei of neuroglial cells.

1. Motor neuron cell body
2. Nucleus
3. Nucleolus
4. Nissl substance
5. Myelinated fibers (purple)
6. Nucleus of neuroglial cell

10. Cerebellum — Luxol Fast Blue, Cresylecht violet — 250×

The molecular layer sparsely occupied by neuroglial cells and the cellular granular layer filled with the same type of supporting cells form two marked divisions. The Purkinje cell layer lies between the two layers.

1. Molecular layer
2. Granular layer
3. Purkinje cell neurons

11. Cerebellum — H & E — 75×

The characteristic appearance of light areas (molecular layer) and dark areas (granular layer) has earned the name arbor vitae ("tree of life"). The pink cerebral cortex lies below.

1. Cerebellum
2. Cerebral cortex
3. Choroid plexus
4. Bone (skull)
5. Granular layer
6. Molecular layer

12. Cerebellum — Bodian silver — 500×

The left half of the field is filled with nucleated cells of the granular layer. Along the inner border of this layer are large Purkinje cells exhibiting axons extending into the molecular layer occupying the right field. Interspersed among the fibers are neuroglial cells.

1. Purkinje cells
2. Axon from Purkinje cell
3. Granular layer
4. Molecular layer
5. Nucleus of neuroglial cell

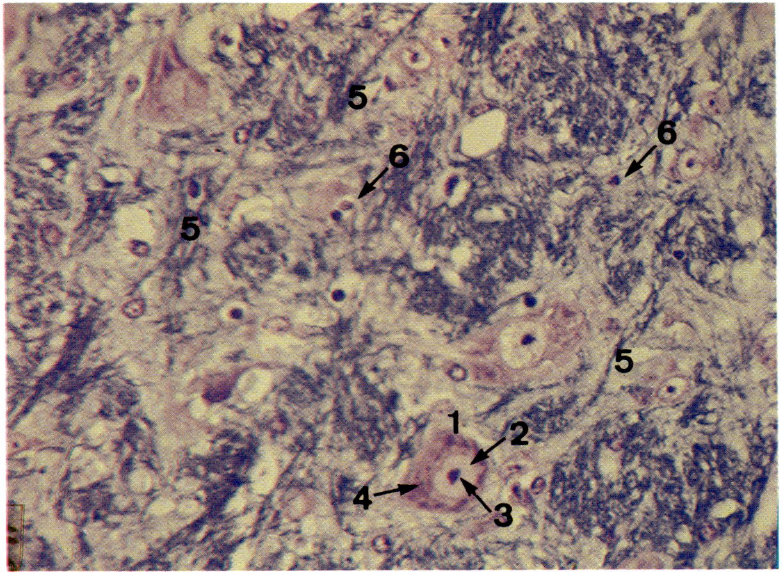

Figure 9

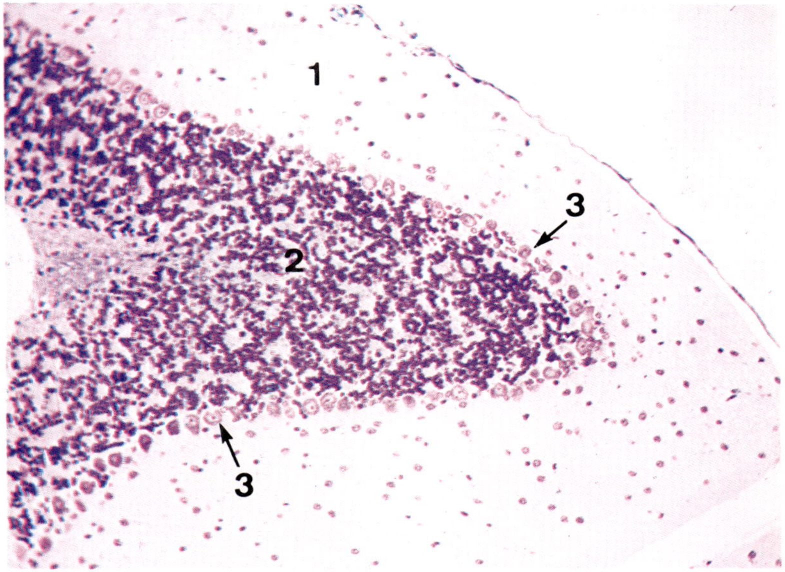

Figure 10

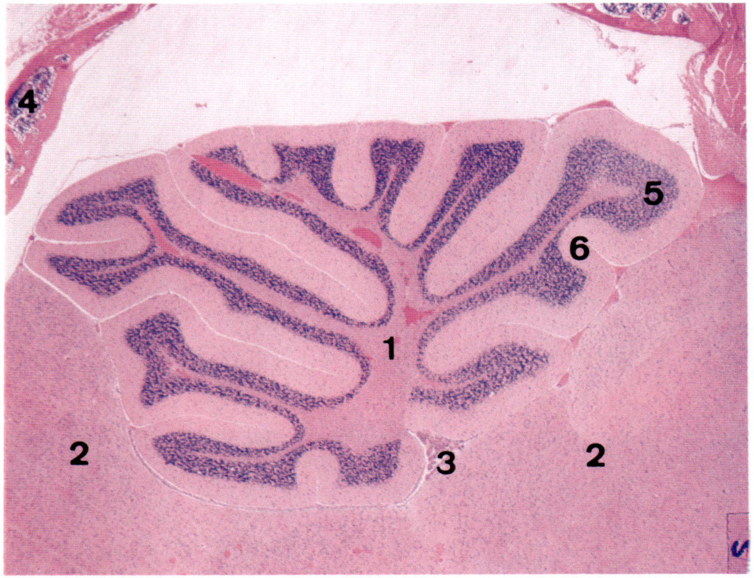

Figure 11

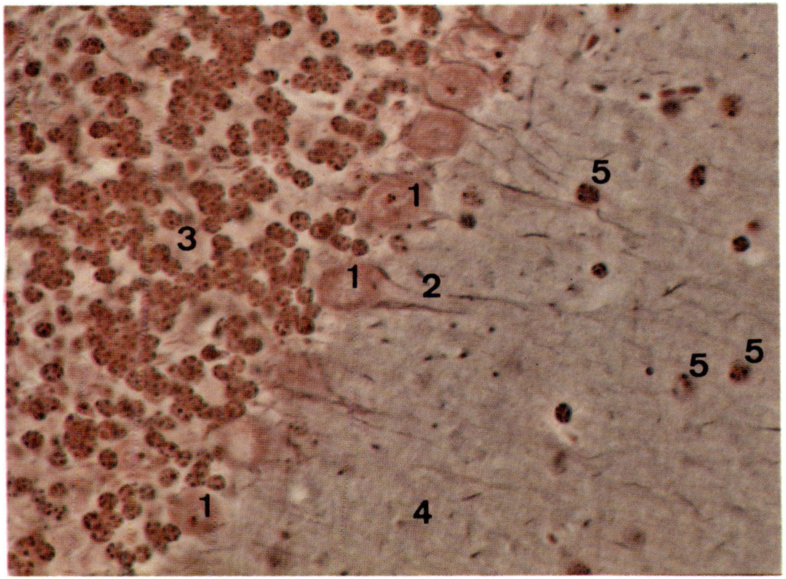

Figure 12

13. Spinal cord — Bodian silver — 75×

In this cross section of the thoracic spinal cord, the posterior horn is in the upper half of the cord, with the anterior horn in the lower half near the vertebral body. A large ganglion from which a nerve trunk extends is seen at lower left.

1. Anterior horn
2. Posterior horn
3. Ganglion
4. Vertebral column
5. Nerve trunk
6. Central canal

14. Spinal cord: Anterior horn cells — Luxol Fast Blue, Nuclear Fast Red — 500×

At this magnification the red nucleoli and chromatin material in the nucleus of large anterior horn motor cells are easily distinguished. A blue-stained capillary is in the right center. One large motor neuron in upper center exhibits an axon and dendrites.

1. Motor cell neuron
2. Neuroglial cell
3. Myelinated nerve fibers
4. Nonmyelinated nerve fibers
5. Blood vessel

15. Spinal cord — Luxol Fast Blue, Nuclear Fast Red — 100×

In this cross section of lumbar spinal cord, the "butterfly" pattern of anterior and posterior horns is apparent, particularly the large red motor cells in the anterior horn. Blue myelinated fibers fill out the surrounding spinal cord. A thin meningeal covering is seen in the space between cord and vertebra.

1. Anterior horn
2. Posterior horn
3. Cerebrospinal canal
4. Myelinated nerve fibers
5. Meninges
6. Pia mater

16. Spinal ganglion — Bodian silver — 250×

Higher magnification enables one to identify the large round ganglion cells and nerve fibers extending toward the lower left.

1. Ganglion nerve cell body
2. Nerve fibers
3. Nerve trunk
4. Skeletal muscle

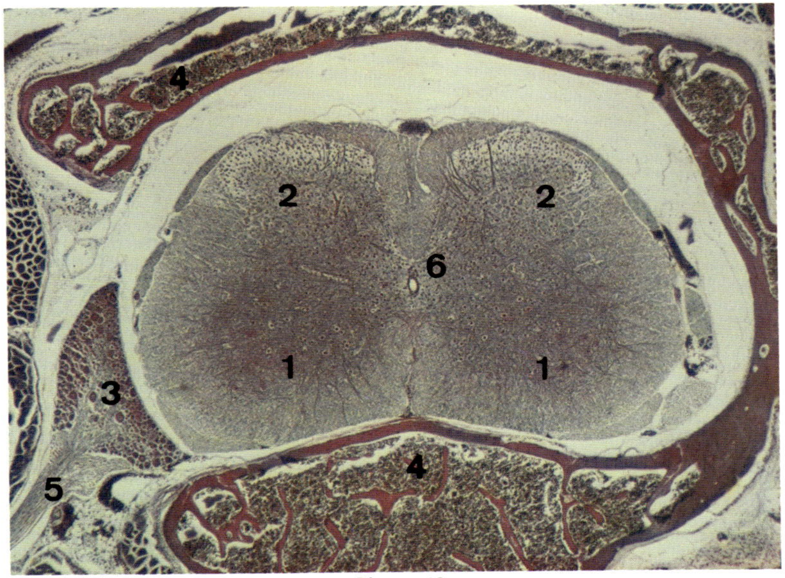

Figure 13

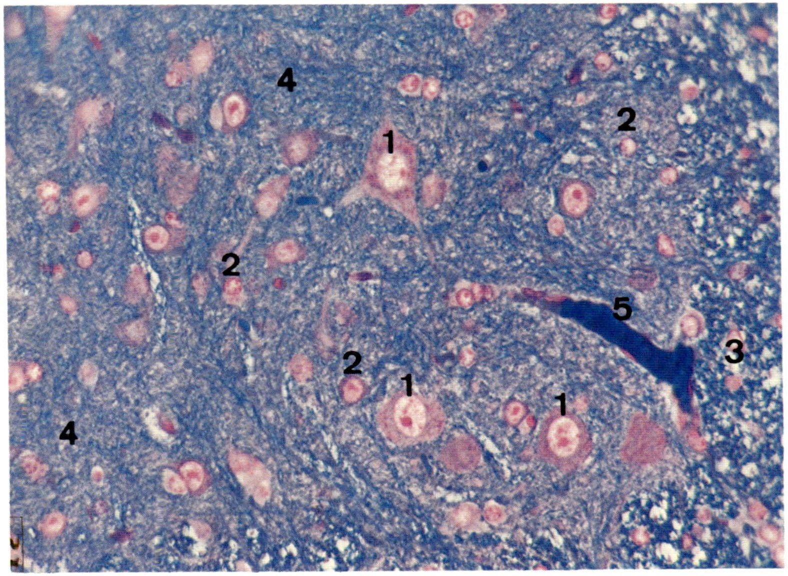

Figure 14

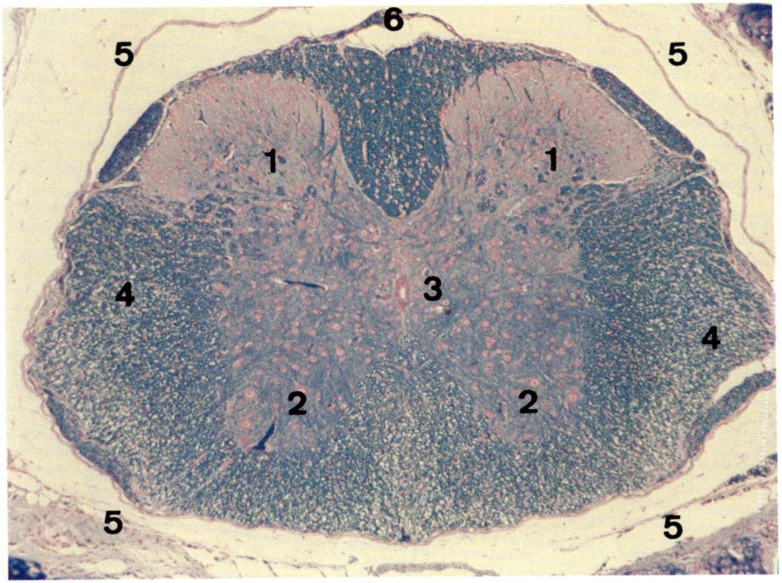

Figure 15

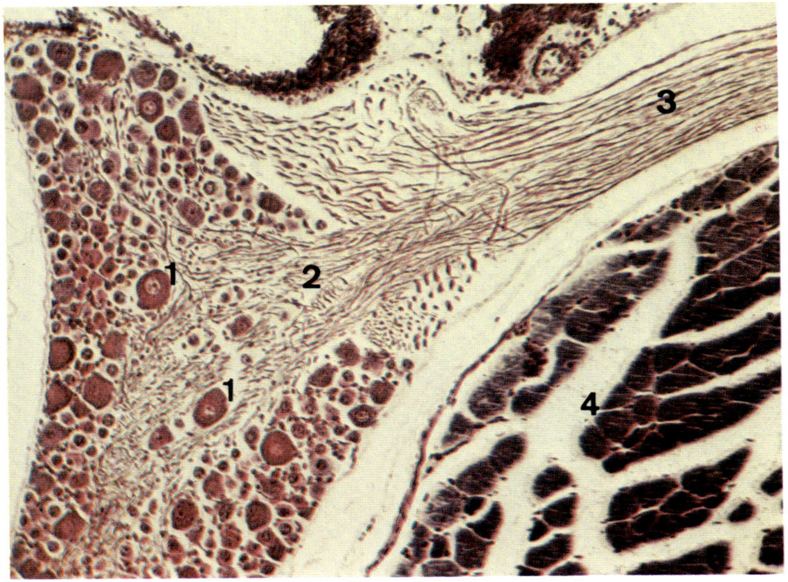

Figure 16

17. Motor end plate — Bodian silver — 500×

Between the bundles of striated muscle fibers are several black nerve fibers terminating on oval, dark brown motor end plates.

1. Nucleus striated muscle
2. Striated (skeletal) muscle
3. Nerve
4. Motor end plate nucleus

18. Motor end plates — Pestronk and Drachman stain — 500×

Acetylcholinesterase appears as an abundant blue mass superimposed on motor end plates. Black nerve fibers enter the end plates from a branch between striated muscle fibers.

1. Nerve fiber
2. Motor end plate
3. Acetylcholinesterase (blue)
4. Striated muscle

19. Pineal gland — H & E — 100×

The pineal gland lies embedded in brain tissue at the junction of the cerebral cortex and cerebellum. Surrounded by blood vessels, it is a highly vascular member of the endocrine system.

1. Pineal gland
2. Cerebral cortex
3. Cerebellum
4. Blood vessel

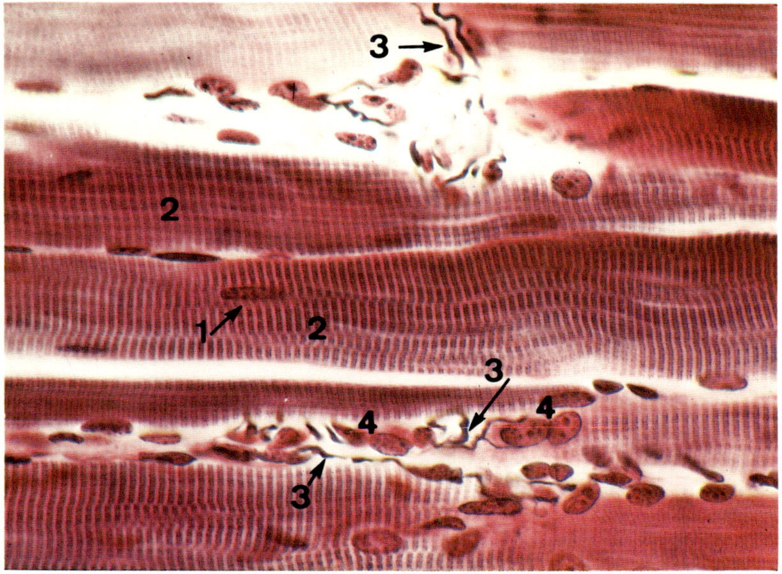

Figure 17

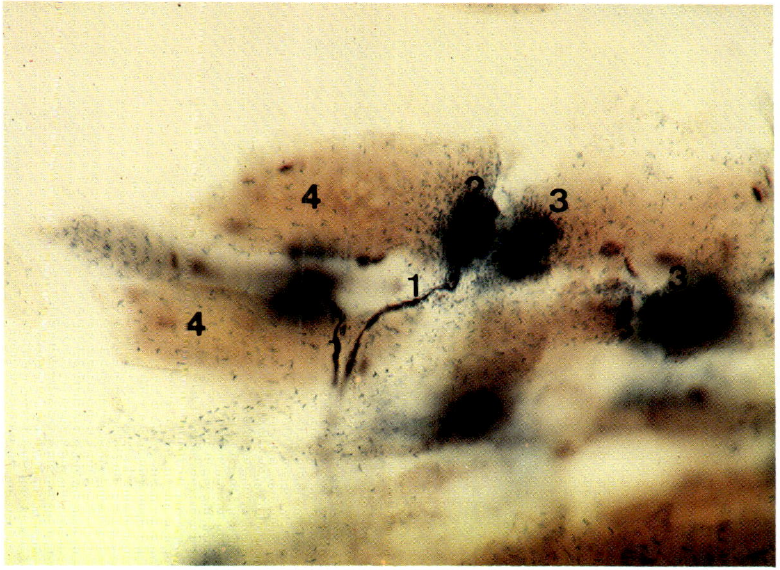

Figure 18

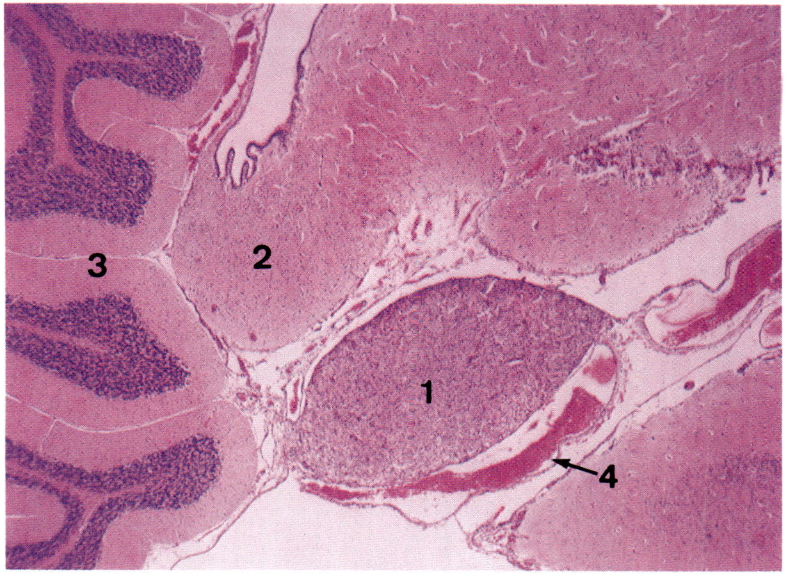

Figure 19

20. Eye H & E 50×

The pink epithelial-lined cornea is the curved tissue on the right extending around half the circumference of the eye. The red central sphere is the lens. The pigmented iris above and below the lens merges into the choroid just posterior to the ciliary body, the small projecting tissue above and below the iris. The retina is the pink band of tissue in the posterior portion of the eye bounded by the ciliary body. The glandular tissue surrounding the eye above posteriorly and below on the left is the Harderian gland (pink).

1. Lens
2. Cornea
3. Iris (pigmented)
4. Retina
5. Choroid (pigmented)
6. Vitreous humor
7. Harderian gland
8. Lacrimal gland

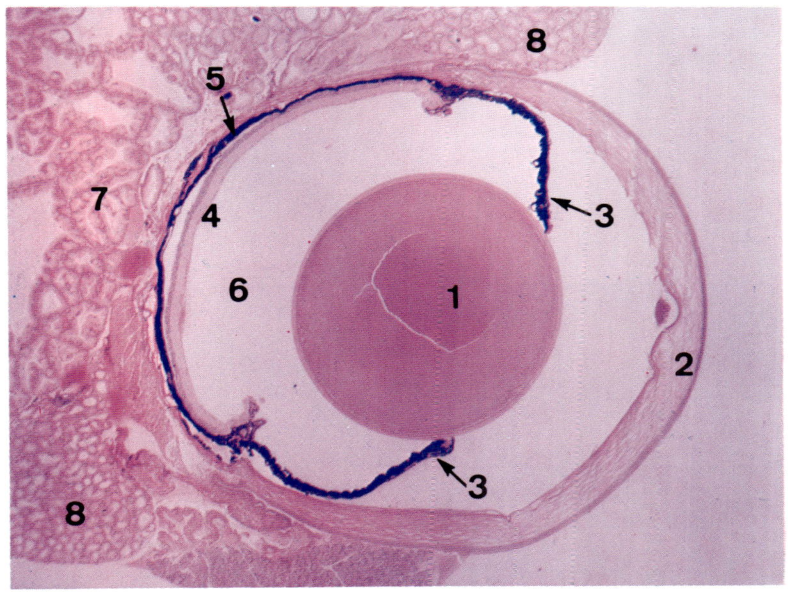

Figure 20

21. Cornea — 1.5 μm Methacrylate, Methylene Blue, Basic fuchsin — 500×

Four distinct layers may be identified. In the outer layer of stratified squamous epithelium, two mitotic figures may be seen, one cell in telophase. Below the epithelium the wide substantia propria layer extends down to a very thin, homogeneous, pink Descemet's membrane, which in turn is immediately adjacent to a posterior endothelium.

1. Stratified squamous epithelium
2. Basal epithelial cell in division
3. Substantia propria
4. Fibroblast
5. Descemet's membrane
6. Endothelium
7. Conjunctiva

22. Eye — H & E — 500×

The large optic nerve fills the central area as it enters the posterior part of the eye ending in immediate contact with the retina. Below, brown fibers of the choroid layer blend with pink fibers of the sclera.

1. Optic nerve
2. Retina
3. Choroid
4. Blood vessel
5. Sclera

23. Eye: Lens, retina, Harderian gland — H & E — 250×

Immediately below the red-stained lens are the deeply pigmented iris on the right and nine-layered retina on the left. Separation of the retina from the pigmented choroid is an artifact. The Harderian gland occupies the lower third of the section.

1. Harderian gland
2. Conjunctiva
3. Ciliary body
4. Iris (pigmented)
5. Retina
6. Lens
7. Choroid (black)

24. Eye: Lens, iris, cornea — H & E — 250×

Across the top of the section, the bow of the lens, covered by a single cell layer of epithelium, contains many nuclei of lens fibers. The wide band of the nine-layered retina extends from the left of the ciliary body in the middle two-thirds of the section. The space below the lens and to the left contains vitreous humor in the living animal. The anterior chamber is the space to the right of the ciliary body, whose pigmented structure is in the center immediately below the lens. Continuation of the heavily pigmented iris extends laterally to the right. The pigmented layer extending to the left is the choroid. The lower third of the section includes the pink sclera immediately subjacent to the choroid, below which is the mucocutaneous epithelium of the conjunctiva, whose upper layer blends with the cornea and whose lower layer continues as the conjunctiva.

1. Lens
2. Ciliary body
3. Iris
4. Retina
5. Sclera
6. Conjunctiva
7. Anterior chamber
8. Posterior chamber
9. Choroid

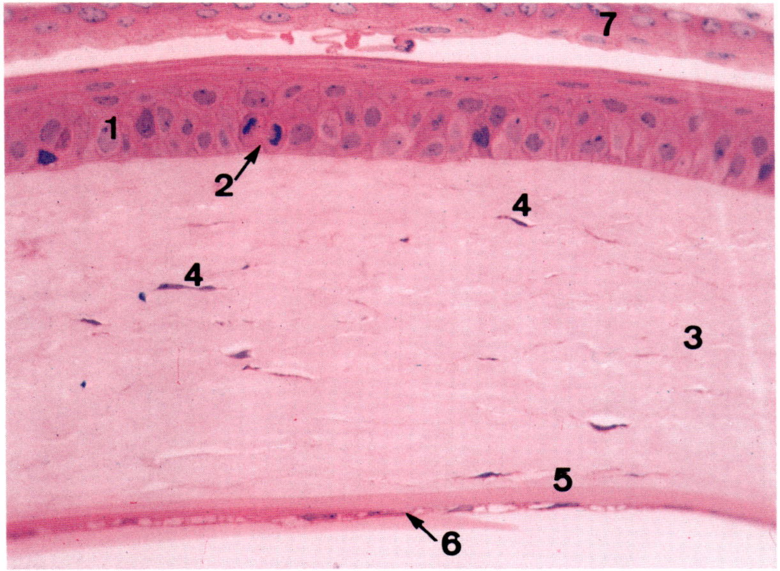

Figure 21

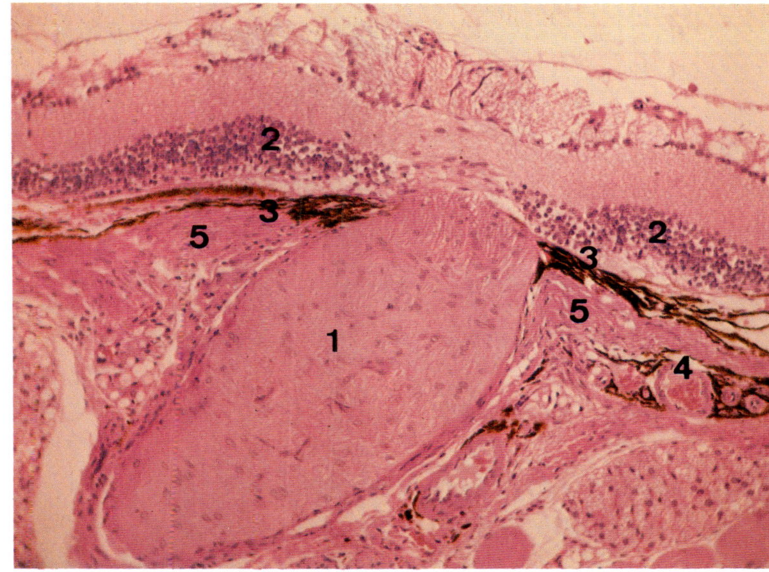

Figure 22

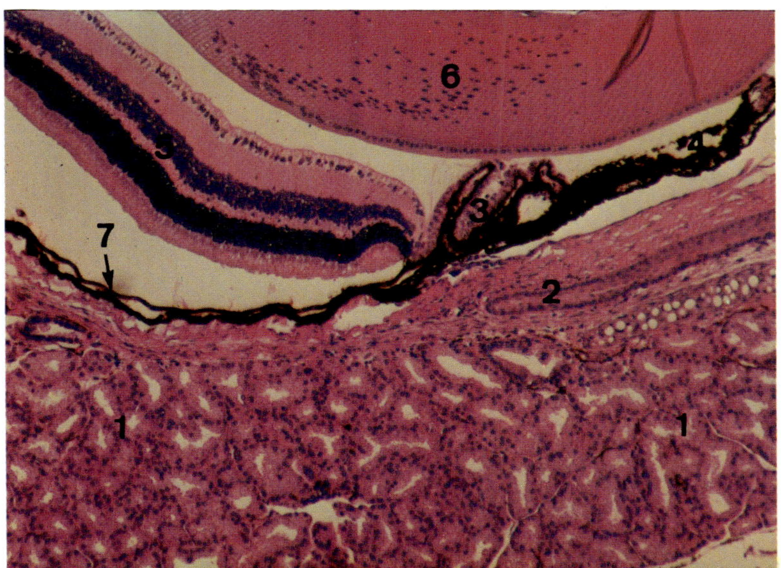

Figure 23

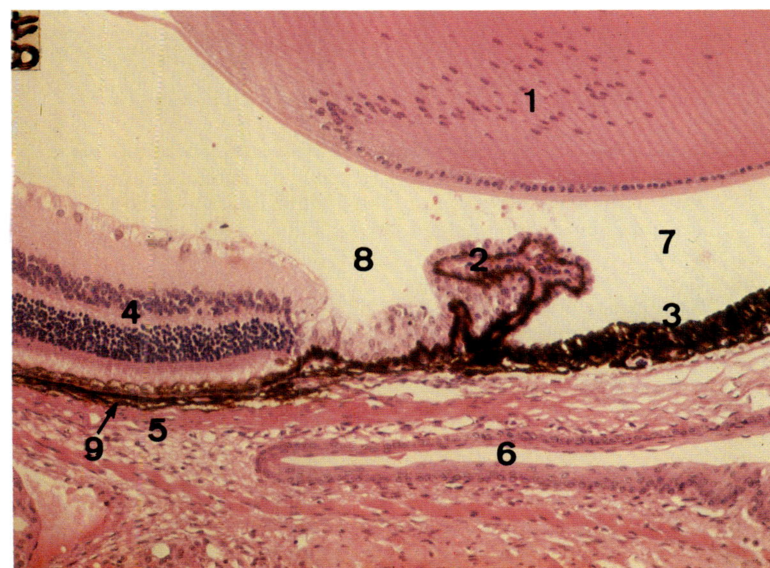

Figure 24

25. Eye: Conjunctiva and cornea PAS 250×

In this section PAS-positive material occurs in the posterior layer (endothelium) of the cornea and at the base of the stratified squamous epithelium. PAS-positive mucus is also seen in cells of the conjunctiva. The sebaceous glands of Meibom are present at lower left.

1. Substantia propria—cornea
2. Epithelium—cornea
3. Conjunctiva—mucous cells (red)
4. Sebaceous gland (Meibomian gland)

26. Eye: Conjunctiva and cornea PAS 275×

The mucus-laden cells of the conjunctiva appear brilliant red, and the endothelium at the posterior cornea as well as some areas of the iris in the bottom of the section are also PAS-positive.

1. Epithelia of cornea
2. Substantia propria
3. Iris
4. Conjunctiva mucous cells (red)

27. Eye: Conjunctiva Heidenhain's Azan 250×

Brown pigment fills the iris at the top of the section. The substantia propria with packed blue collagen fibers dominates the cornea. Below is the conjunctiva with blue-staining mucoid cells in its epithelium. Striated muscle fibers appear pink at lower right and Meibomian sebaceous glands appear at lower left.

1. Conjunctiva mucous cells (light blue)
2. Meibomian (sebaceous) gland
3. Striated muscle fibers
4. Cornea
5. Pigmented iris

28. Eye: Lacrimal and Harderian glands Masson stain 250×

In the left lower third of the section is a portion of the lacrimal gland. Its basophilic glandular appearance resembles the serous-type salivary gland. The pink glandular tissue on the right is part of the Harderian gland.

1. Striated muscle
2. Harderian gland
3. Lacrimal gland
4. Ciliary body
5. Retina
6. Choroid (black)
7. Lens
8. Iris (pigmented)
9. Conjunctiva (eyelid)
10. Vitreous humor space

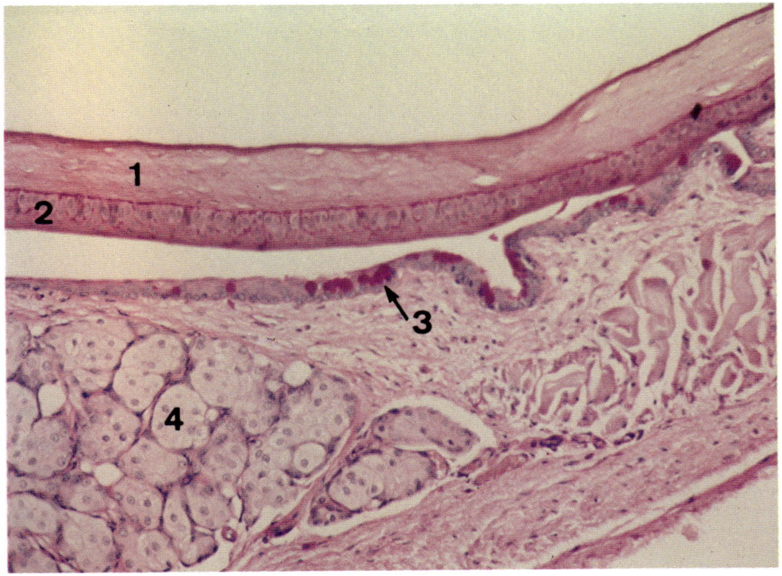

Figure 25

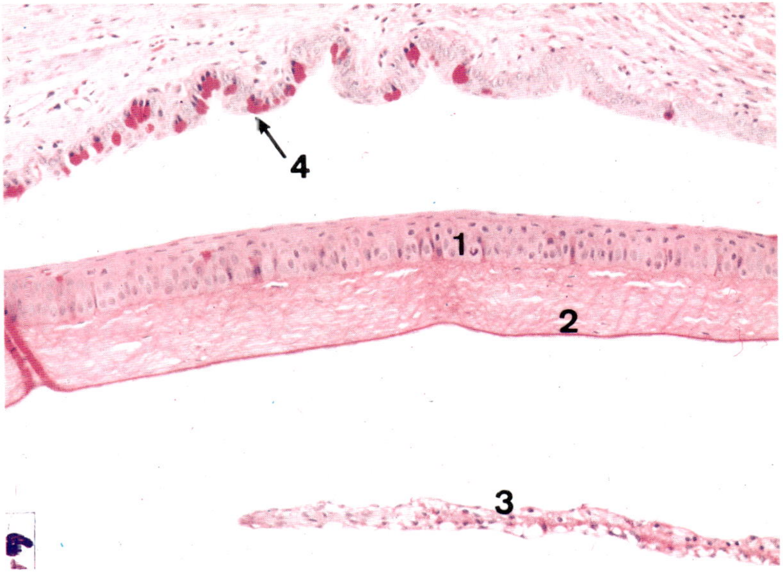

Figure 26

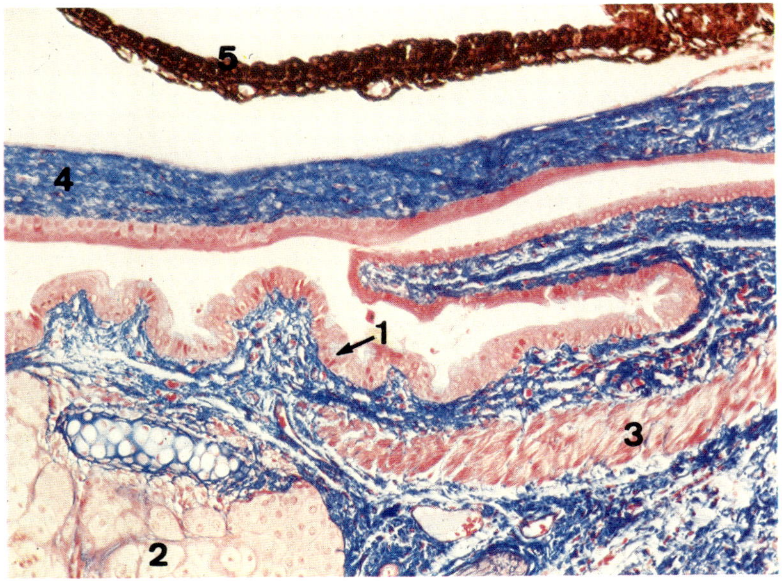

Figure 27

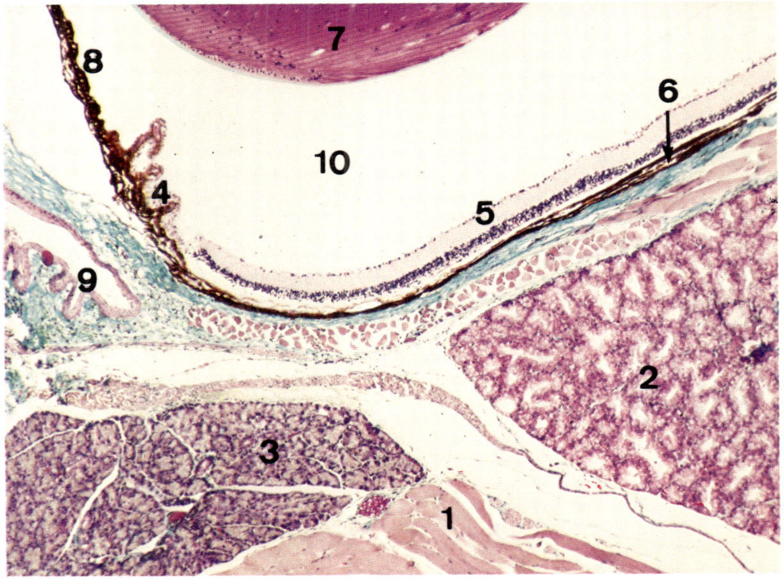

Figure 28

29. Ear: Organ of Corti H & E 75×

This is a section cut through the bony cochlea to illustrate the delicate structures identified in the captions associated with the photomicrograph.

1. Limbus lamina spiralis
2. Vestibular membrane (Reissner's)
3. Scala vestibuli
4. Cochlear duct
5. Basilar membrane
6. Internal spiral sulcus
7. Spiral ligament
8. Tectorial membrane
9. Scala tympani
10. Stria vascularis
11. Hair cells

30. Ear: Organ of Corti Heidenhain's Azan 250×

At each end of the center structure the blue tectorial membrane overhangs the tiny hair cells resting on the basilar membrane. The thin structure separating the cochlear duct from the scala vestibuli is the vestibular or Reissner's membrane. The scala tympani is the space below the basilar membrane.

1. Tectorial membrane
2. Vestibular membrane (Reissner's)
3. Limbus spiralis
4. Scala tympani
5. Hair cells
6. Cochlear duct
7. Scala vestibuli

31. External ear (pinna) H & E 50×

Stratified squamous epithelium of epidermis forms the outer border. The pink area enclosing spaces in the center is elastic cartilage. The pink connective tissue between outer epidermis and layer of cartilage is dermis, interspersed with groups of large, pale sebaceous gland cells. A few spherical cross sections of hair follicles are found in the dermis.

1. Stratified squamous epithelium
2. Dermis
3. Sebaceous gland
4. Hair follicle
5. Fat cells
6. Elastic cartilage
7. Striated muscle

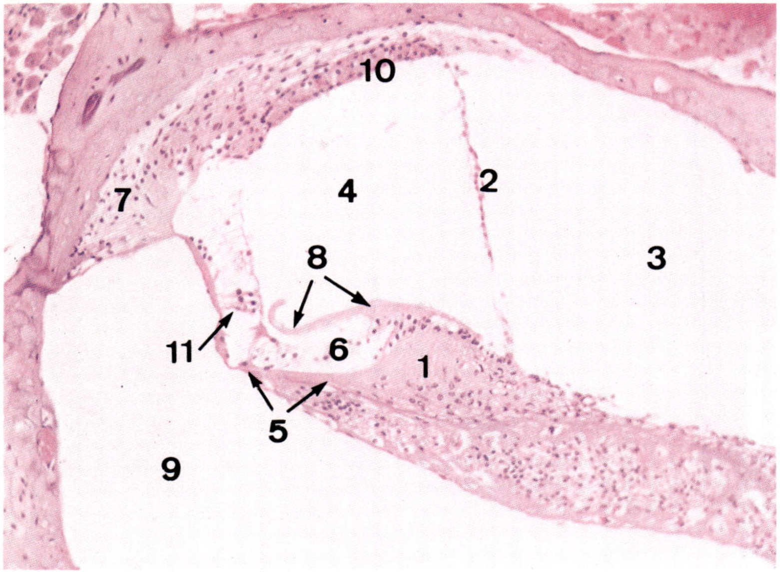

Figure 29

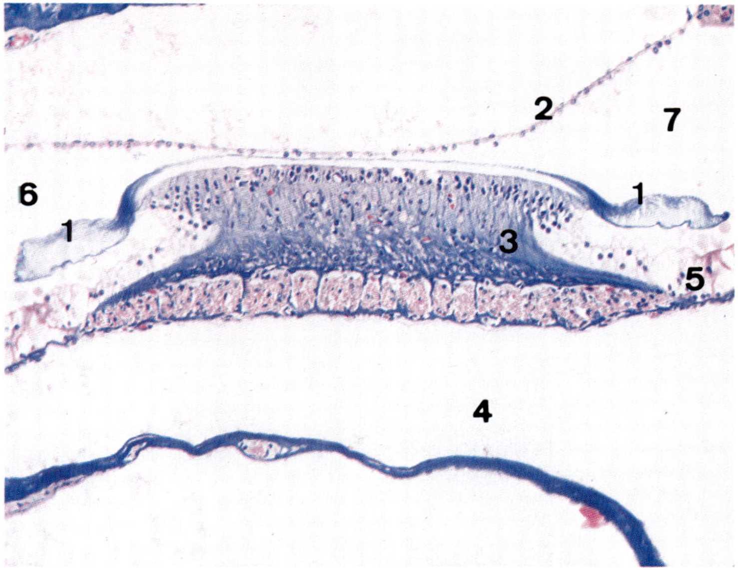

Figure 30

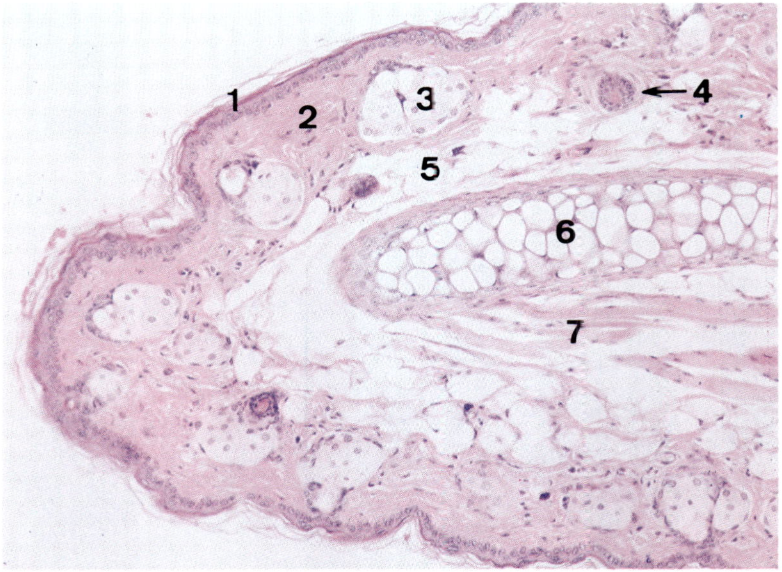

Figure 31

32. Pituitary gland — Performic acid, Aldehyde–fuchsin — 200×

The three lobes of the pituitary are clearly seen. The posterior pituitary has reacted positively to the stain, which demonstrates the presence of neurosecretory substance.

1. Anterior lobe
2. Intermediate lobe
3. Posterior lobe

33. Pituitary gland: Anterior lobe — Aldehyde–fuchsin, Martin–Mallory stain — 250×

Red acidophils predominate in this section, with an occasional purple beta cell (gonadotroph) and many chromophobes (gray). Two arteries with distinctive internal elastic membranes (purple) are present in upper right.

1. Aldehyde-positive basophil
2. Acidophil
3. Chromophobe
4. Artery
5. Collagen fibers (blue)

34. Pituitary gland: Hypophysis — Masson stain — 75×

The pituitary gland lies anatomically within the bony sella turcica (green), with marrow cells within its bony divisions. The red-staining anterior lobe is separated by a cleft from the narrow intermediate lobe and the lighter pink posterior lobe (neurohypophysis).

1. Anterior lobe
2. Intermediate lobe
3. Posterior lobe
4. Bone marrow
5. Bone
6. Striated muscle

35. Pituitary gland: Anterior lobe — Aldehyde–fuchsin, Acid fuchsin, Orange G — 500×

Aldehyde–fuchsin-positive gonadotrophs (purple) are easily differentiated among red acidophils and gray chromophobes.

1. Aldehyde–fuchsin-positive beta cells
2. Acidophils
3. Chromophobes

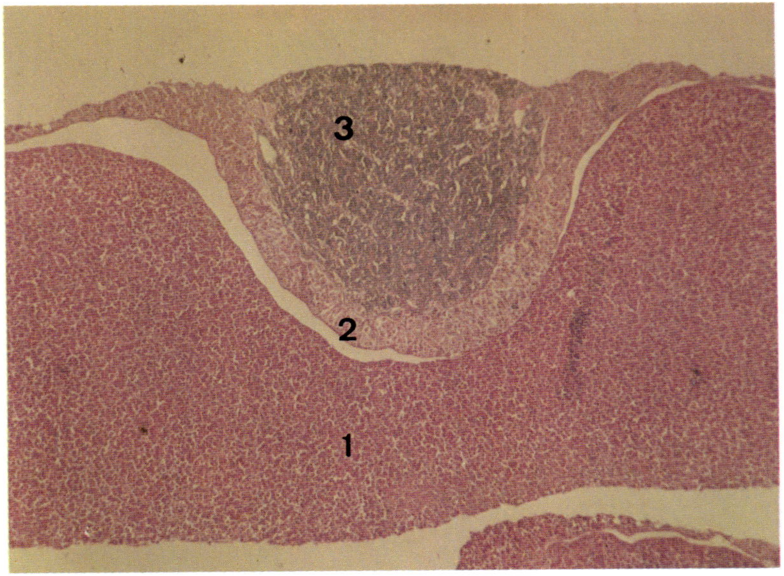

Figure 32

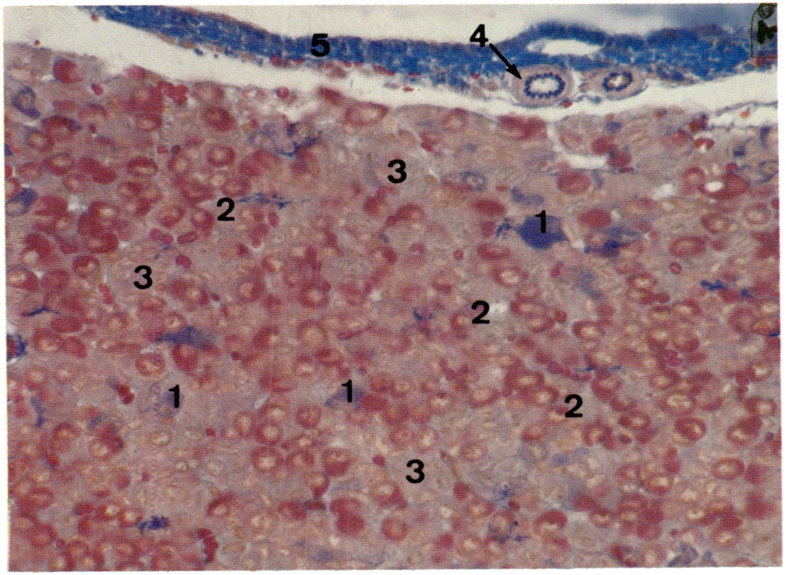

Figure 33

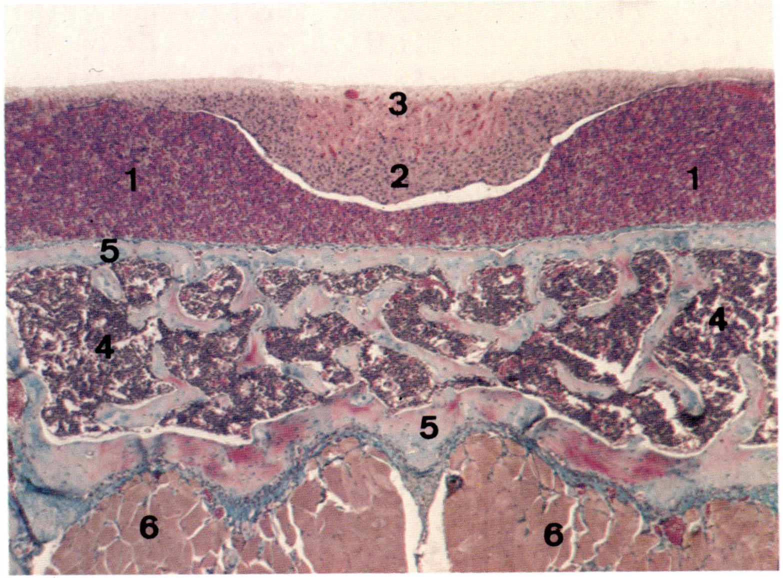

Figure 34

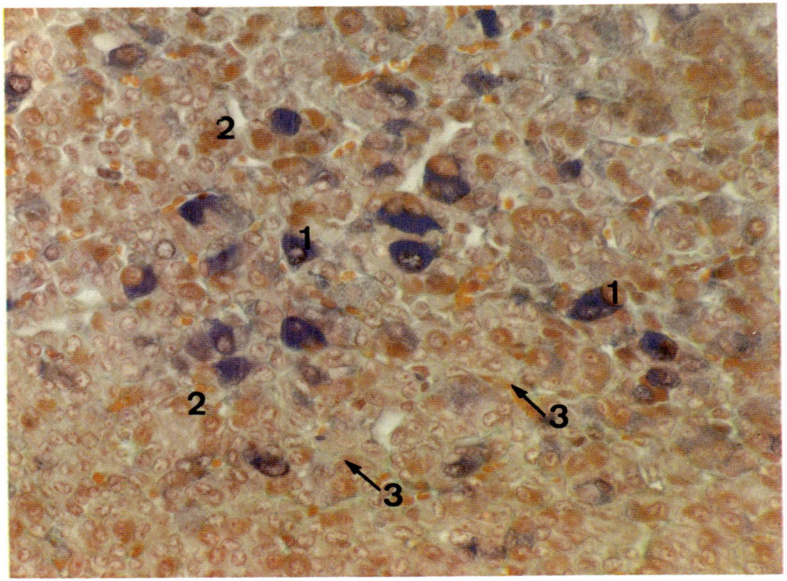

Figure 35

36. Adrenal gland H & E 100×

Division between cortex and medulla is apparent. In the cortex the outer zona glomerulosa lies immediately below the capsule and is adjacent to the zona fasciculata. The zona fasciculata is wide and extends to the next zone, which is the X zone, also known as the "juvenile zone." The zona reticularis does not appear in this section. The medulla (center) contains several blood vessels.

1. Zona glomerulosa
2. Zona fasciculata
3. X zone
4. Medulla
5. Blood vessel
6. Kidney

37. Adrenal gland H & E 200×

The three zones of the cortex and the medulla can easily be identified.

1. Capsule
2. Zona glomerulosa
3. Zona fasciculata
4. X zone
5. Medulla

38. Adrenal gland PAS 50×

At low magnification the extent of "brown degeneration" can easily be seen. The presence of lipid in this substance results in a positive reaction (red) with periodic acid–Schiff reagent concentrated between the zona fasciculata (cortex) and margin of the medulla.

1. Zona fasciculata (cortex)
2. "Brown degeneration" pigment
3. Medulla
4. Kidney cortex
5. Glomerulus

Sample from $1\frac{1}{2}$-yr-old mouse. Brown degeneration increases as mouse ages.

39. Adrenal gland H & E 100×

A large amount of cortical "brown degeneration" pigment fills the central and right field. A small area of medulla is seen at lower right.

1. Zona glomerulosa
2. Zona fasciculata
3. "Brown degeneration" pigment
4. Medulla

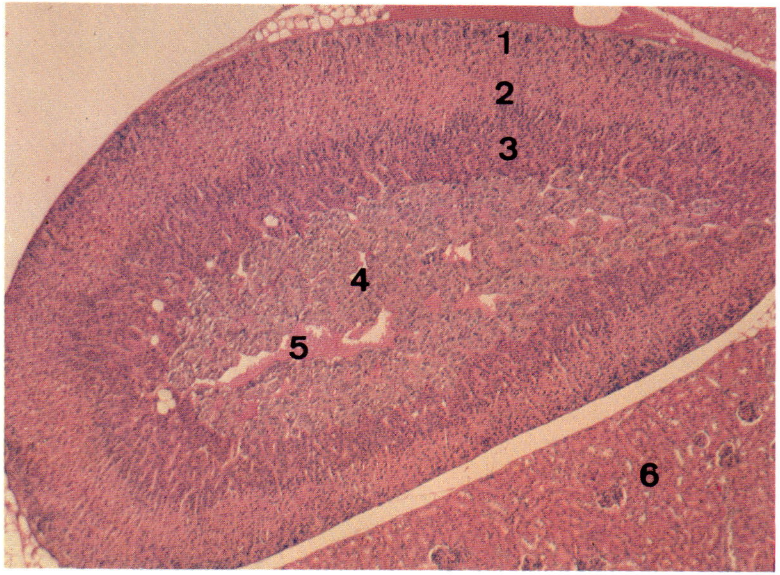

Figure 36

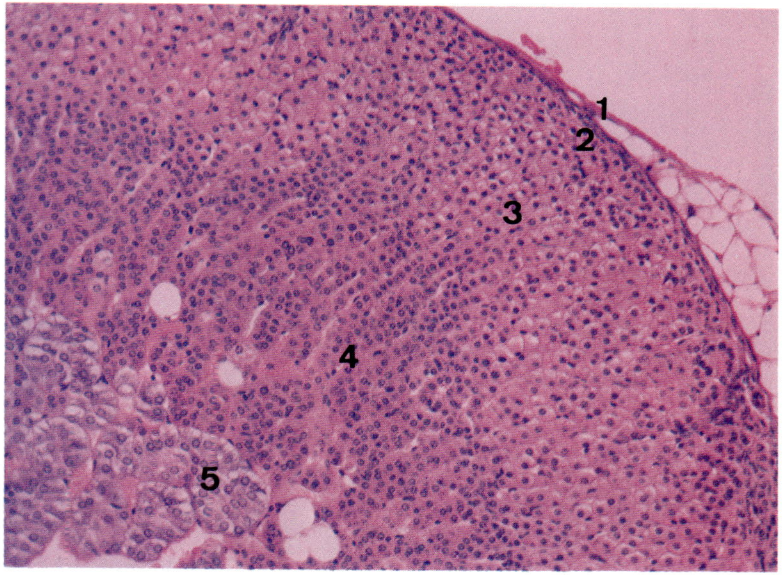

Figure 37

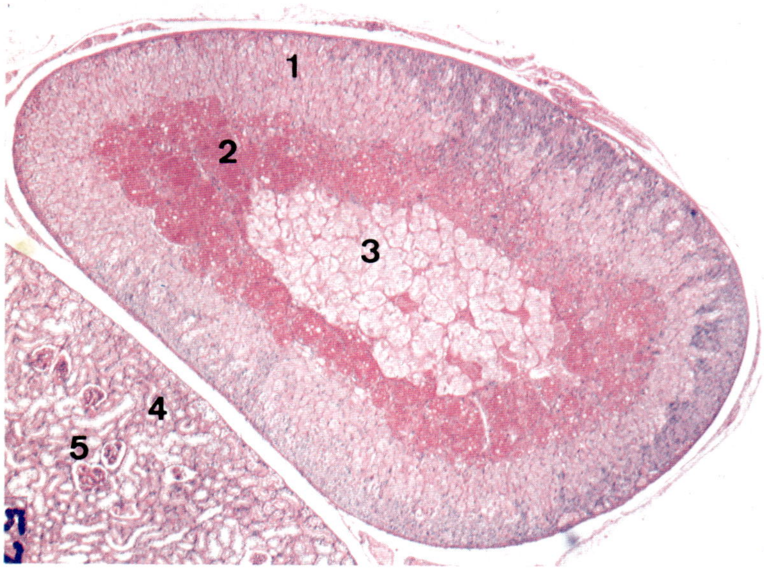

Figure 38

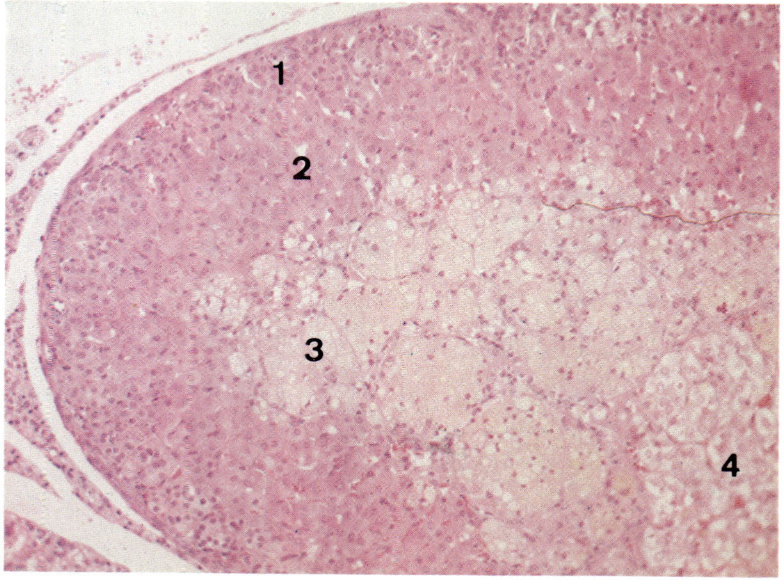

Figure 39

40. Thyroid and parathyroid glands PAS 50×

The anatomical relationships of the thyroid gland, trachea, parathyroid gland, and esophagus are seen in this low-power section. The cartilaginous, ringed trachea is bound by the lateral lobes of the thyroid gland, whose brightly stained red colloid is apparent. The isthmus lobe, which lies on the anterior surface of the trachea, is not present in this section. Posterior to the trachea is the esophagus.

1. Lumen of trachea
2. Lumen of esophagus
3. Hyaline cartilage
4. Mucous glands of trachea
5. Stratified squamous epithelium—esophagus
6. Striated muscle
7. Thyroid gland
8. Parathyroid gland
9. Striated muscle
10. Epithelium—trachea

41. Thyroid and parathyroid glands H & E 500×

Cuboidal epithelium lines the colloid-filled follicles of the thyroid gland. Chief and oxyphil cells of the parathyroid gland appear as light and dark blue-stained cells.

1. Colloid within follicle—thyroid
2. Follicular cells—thyroid
3. Parathyroid gland

42. Thyroid and parathyroid glands H & E 250×

Cuboidal epithelium lines the thyroid follicles, which are filled with pink colloid. The dark-purple-stained parathyroid gland is penetrated by several blood vessels. Striated muscle of the esophagus appears at upper left.

1. Thyroid follicles
2. Parathyroid gland
3. Striated muscle

43. Parathyroid gland H & E 250×

Two cell types can be identified: chief cells (ovoid, very dark nuclei, eosinophilic cytoplasm) and principal cells (large vesicular nucleus and smaller amounts of basophilic cytoplasm).

1. Chief cells
2. Oxyphil cells

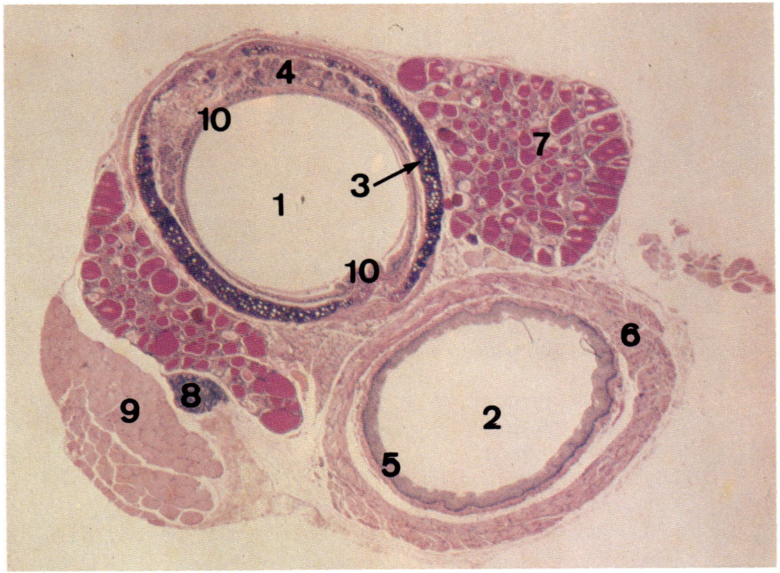

Figure 40

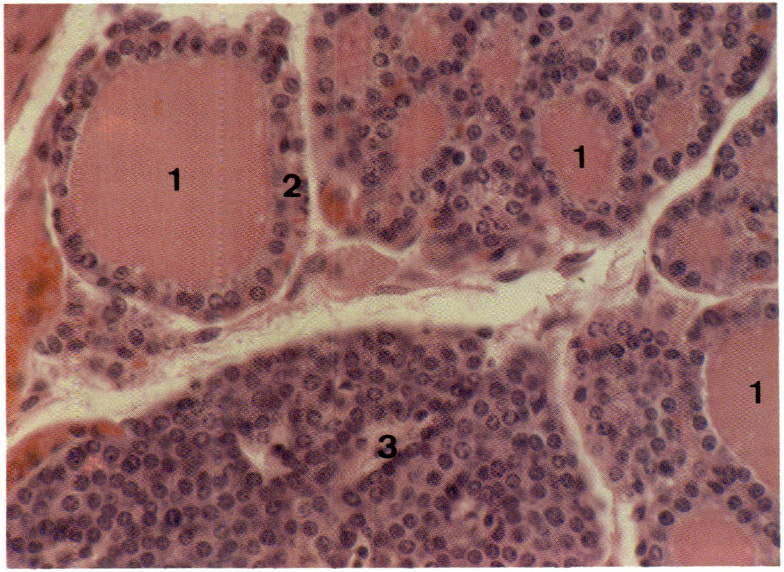

Figure 41

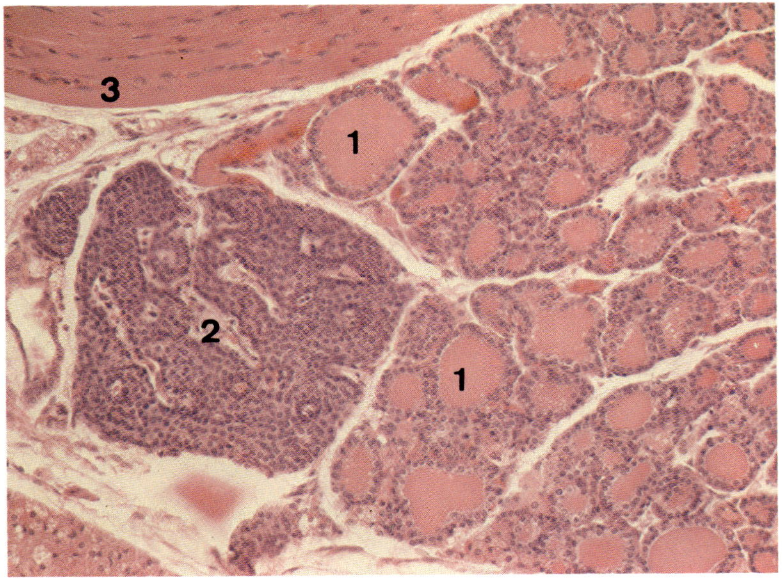

Figure 42

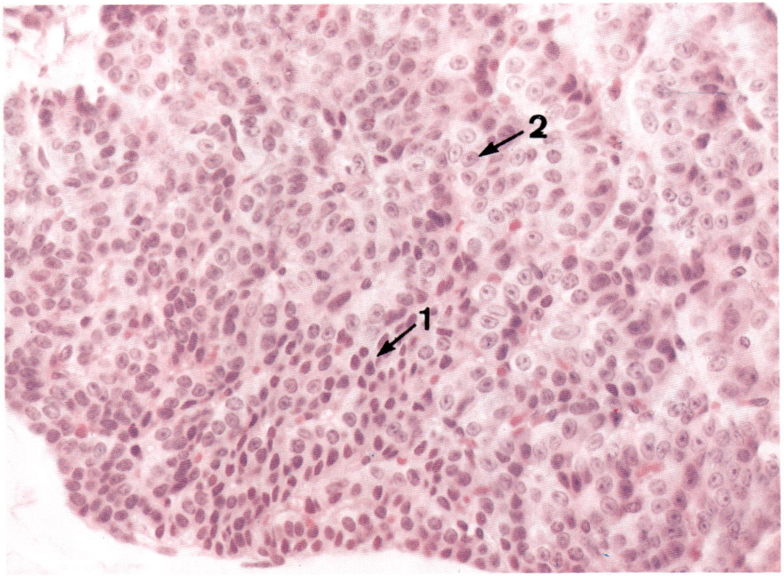

Figure 43

44. Ovary H & E 50×

The ovary is completely surrounded by a capsule of loose connective tissue. Beneath the capsule is the "germinal" epithelium of cuboidal cells resting on a basement membrane. Adipose tissue around the ovary is seen at upper right. A small portion of uterine tube appears at upper left. In the ovary several small primordial follicles can be seen adjacent to the capsule at lower left, while a growing follicle with follicular fluid in the antrum is at upper right. Immediately below this follicle are two corpora lutea with their characteristic pink luteal cells and capillaries.

1. Growing follicle
2. Corpus luteum
3. Primordial follicle
4. Germinal epithelium
5. Fat cells
6. Capsule

45. Ovary H & E 500×

The "germinal" epithelium forms the upper boundary of the ovary, beneath which are several primordial cells. Two oocytes are below the primordial cells, left and right. A larger growing follicle occupies the center. The mature ovum at lower center consists of the nucleus and cytoplasm of the ovum, and the antrum (space), containing follicular fluid, is surrounded by a ring of follicular cells.

1. Germinal epithelium
2. Primordial follicle
3. Growing follicle
4. Nucleus of ripening follicle } Ovum
5. Cytoplasm of ripening follicle } Ovum
6. Follicular cells
7. Interstitial cells

46. Ovary H & E 500×

A large maturing follicle with ovum in the center is seen in the middle of this section. Surrounding the ovum is a pink, oval ring, the zona pellucida, outside of which are several rows of follicular cells. The space at left center is filled with follicular fluid. One follicular cell is dividing. Outside the ring of follicular cells are cells forming the theca. Occasionally one can differentiate a theca interna separated from theca externa by capillaries. Part of a growing follicle is seen above the mature ovum.

1. Mature follicle
2. Ovum nucleus
3. Follicle cell in mitosis
4. Follicular fluid
5. Growing follicl
6. Blood vessel
7. Primordial follicle
8. Theca
9. Zona pellucida

47. Vagina H & E 50×

The external vaginal opening is formed between layers of stratified squamous epithelium. Sebaceous glands lie on both sides, surrounded by smooth muscle. Skin with hair follicles and sebaceous glands merges with the vaginal epithelium.

1. Vaginal opening
2. Stratified squamous epithelium
3. Urethra
4. Skin
5. Clitoral gland
6. Sebaceous gland
7. Fat cells

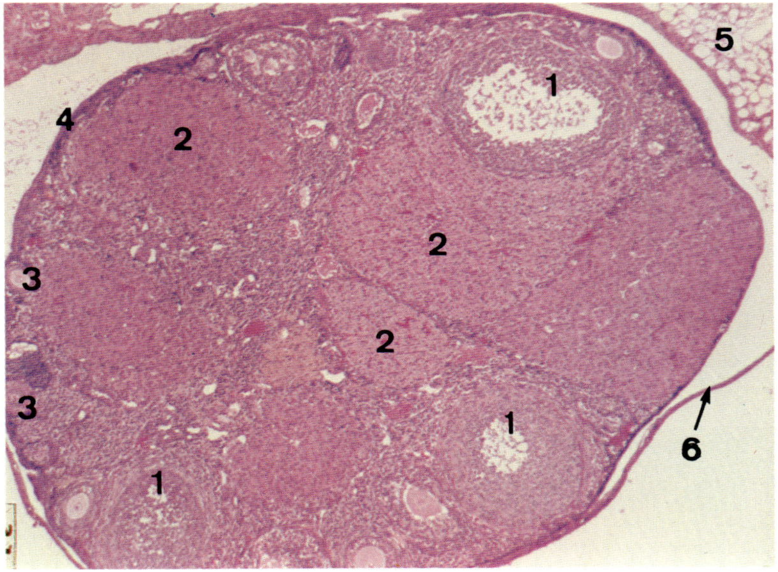

Figure 44

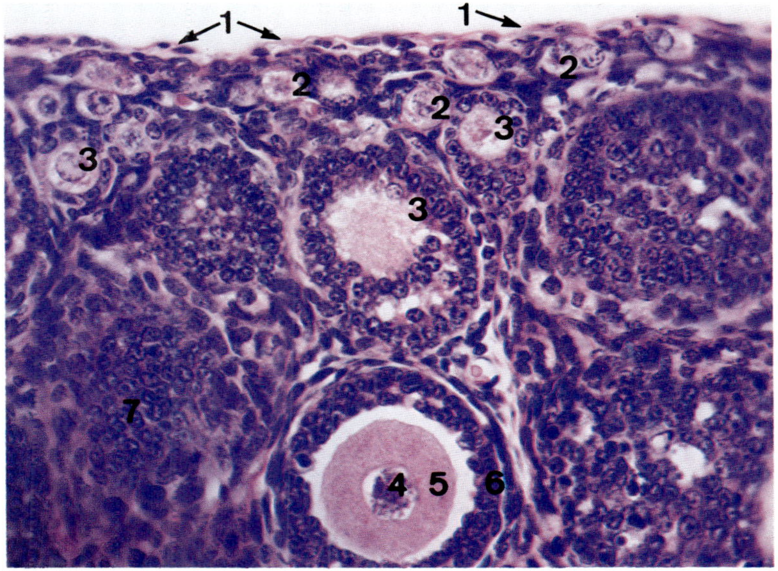

Figure 45

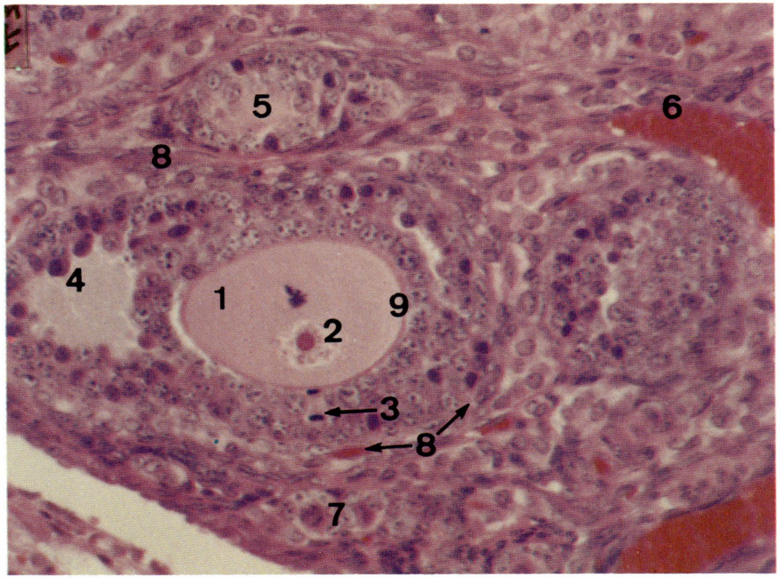

Figure 46

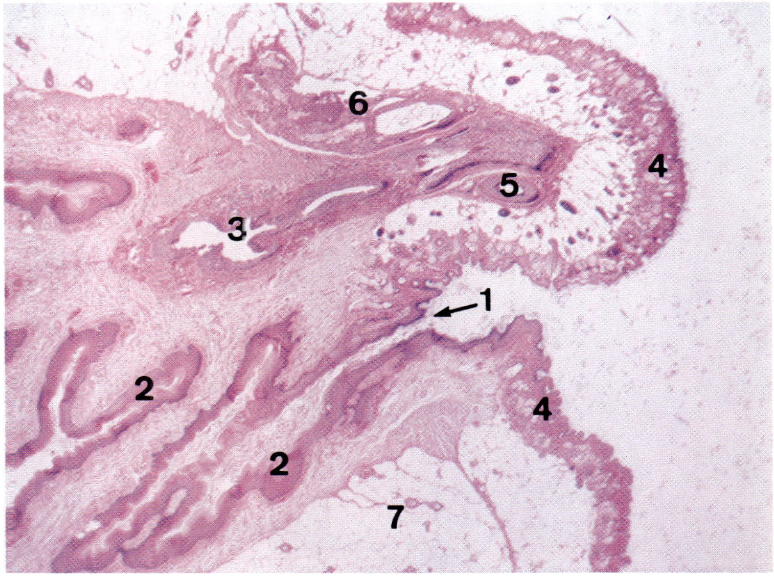

Figure 47

48. Uterus PAS 150×

Uterine glands with PAS-positive (red) glycogen are easily recognizable in the endometrium occupying the right half of the section. The myometrium of circular and longitudinal smooth muscle is seen at center and to the left bounded by the serosa.

1. Uterine glands (glycogen, red)
2. Lumen of uterus
3. Simple columnar epithelium
4. Endometrium
5. Myometrium
6. Blood vessels
7. Serosa

49. Fimbria of oviduct Heidenhain's Azan 500×

Ciliated cuboidal epithelium lines this structure, with a pronounced red striated border appearing as a base from which cilia extend. Blue collagen fibers are part of the connective tissue supporting the epithelium.

1. Ciliated columnar epithelium
2. Lumen of oviduct
3. Striated border (red)

50. Placenta and embryo H & E 50×

In this section an embryo is located in the center, surrounded by a very thin amnion, outside of which is the visceral yolk sac. The fetal placenta is seen at the bottom of the photo while the maternal placenta (decidua basalis) occupies the upper third of the section.

1. Embryo
2. Visceral yolk sac
3. Decidua basalis
4. Endometrium
5. Blood island
6. Amnion (very thin)
7. Decidua capsularis
8. Metrial gland

51. Placenta H & E 200×

The upper two-thirds of the section is the decidua basalis with maternal blood cells within the blood vessels. The middle section contains a few giant cells separated by spaces (labyrinths). Below is the fetal placenta, whose blood vessels are filled with embryonic blood cells.

1. Giant-cell area
2. Fetal placenta
3. Decidua basalis
4. Labyrinth
5. Maternal blood cells
6. Embryonic blood cells

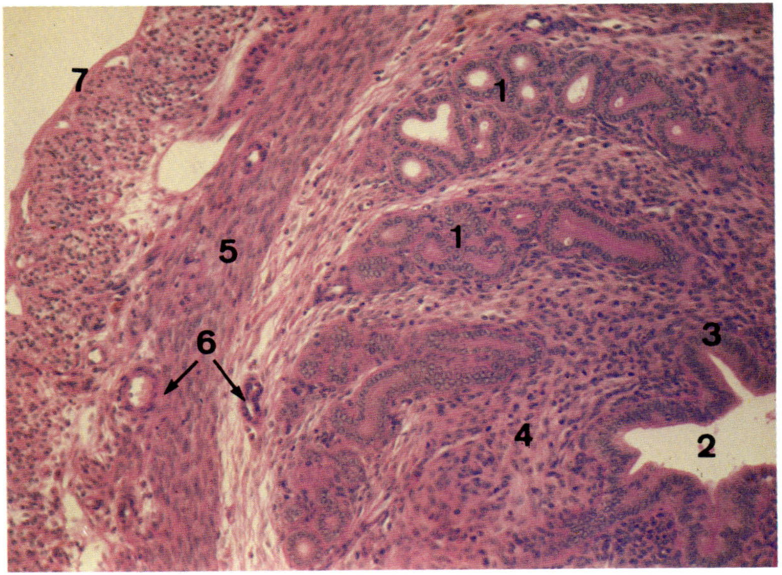

Figure 48

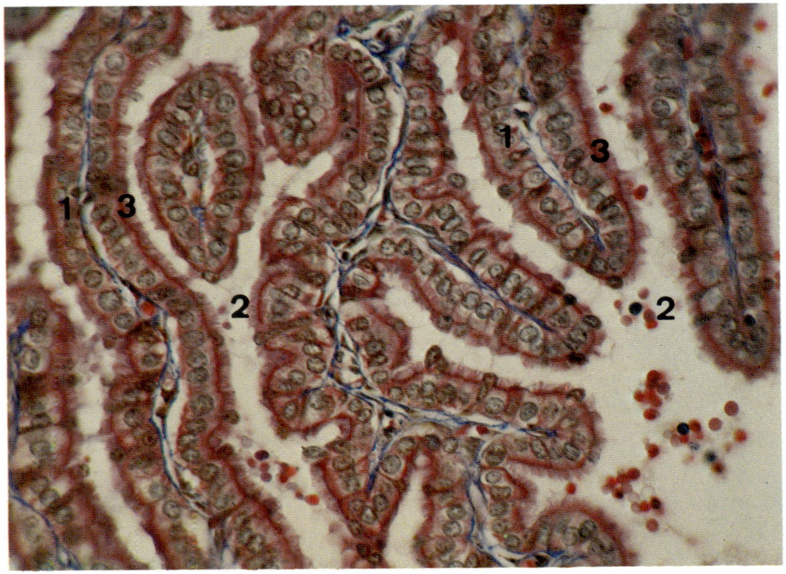

Figure 49

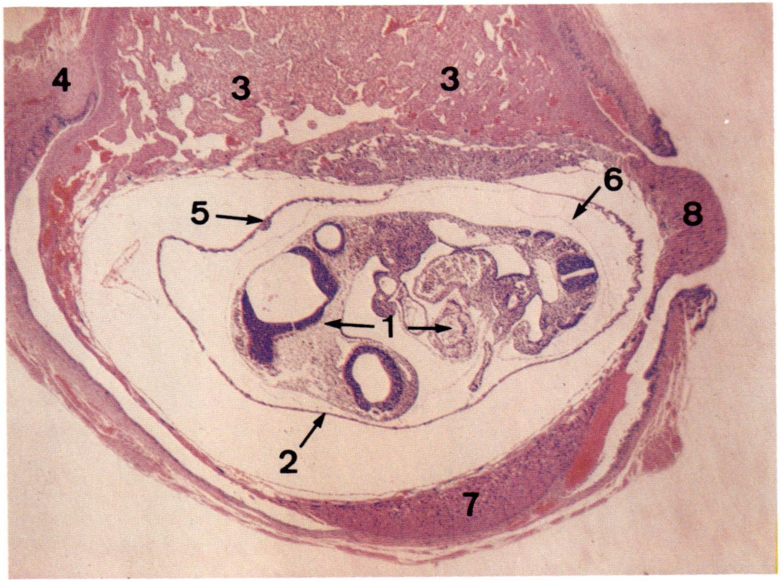

Figure 50

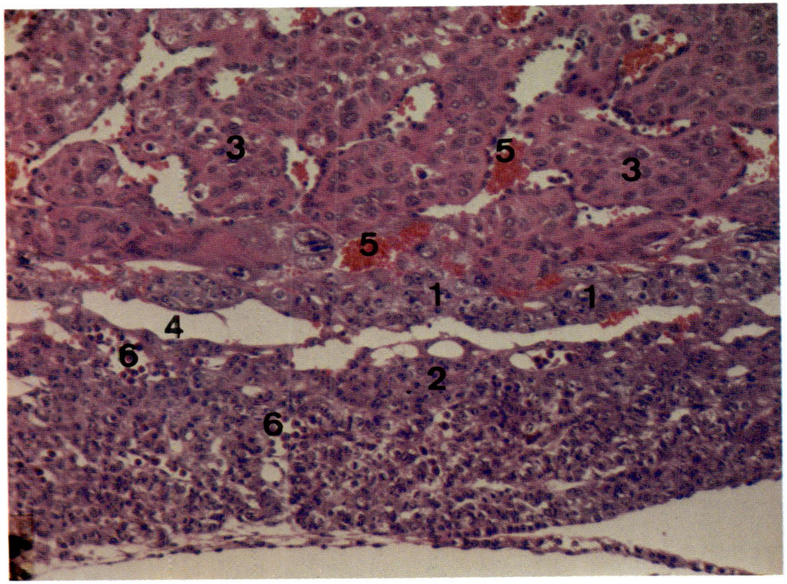

Figure 51

52. Placenta: Maternal side H & E 50×

The maternal structures are identified as the myometrium on the right adjoining the endometrium to the left, which in turn is adjacent to the very vascular decidua basalis, below which is the fetal placenta.

1. Myometrium
2. Endometrium
3. Decidua basalis
4. Fetal placenta

53. Vagina: Diestrus H & E 250×

The diestrus phase is one in which the epithelium begins to rebuild, the stratum germinativum increasing to a thickness of several cell layers. There is also an abundance of polymorphonuclear leukocytes moving from the epithelium out into the lumen.

1. Stratified squamous epithelium (includes proliferating stratum germinativum)
2. Polymorphonuclear leukocytes
3. Connective tissue and smooth muscle

54. Vagina: Proestrus H & E 450×

A thin, red cornified layer appears just below the stratum granulosum during the beginning of proestrus. A few mitotic figures appear.

1. Granulosa layer of epithelium beginning cornification
2. Thickening stratified squamous epithelium
3. Mitotic figures

55. Vagina: Proestrus/estrus H & E 250×

Occasionally, under the influence of progesterone, a layer of mucous cells forms on the outer stratum granulosum. They slough into the lumen, followed by increased thickening of the cornified layer.

1. Mucous cells
2. Thin red line of cornification
3. Stratified squamous epithelium

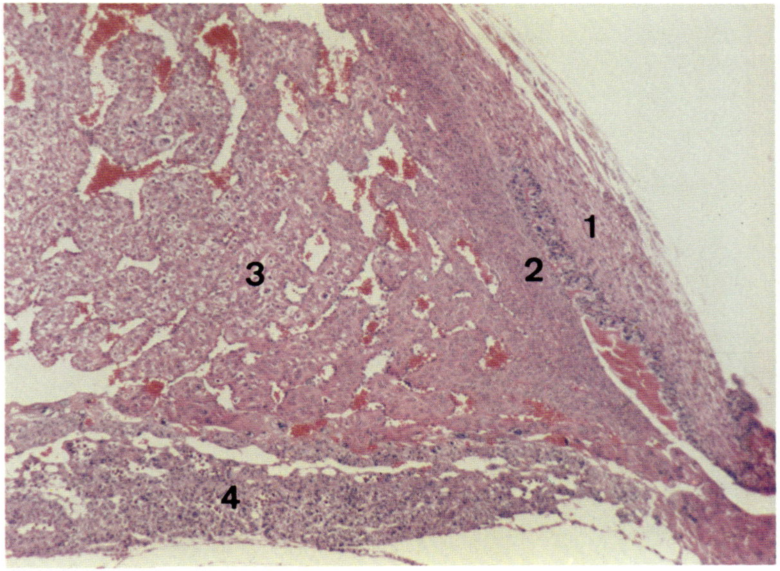

Figure 52

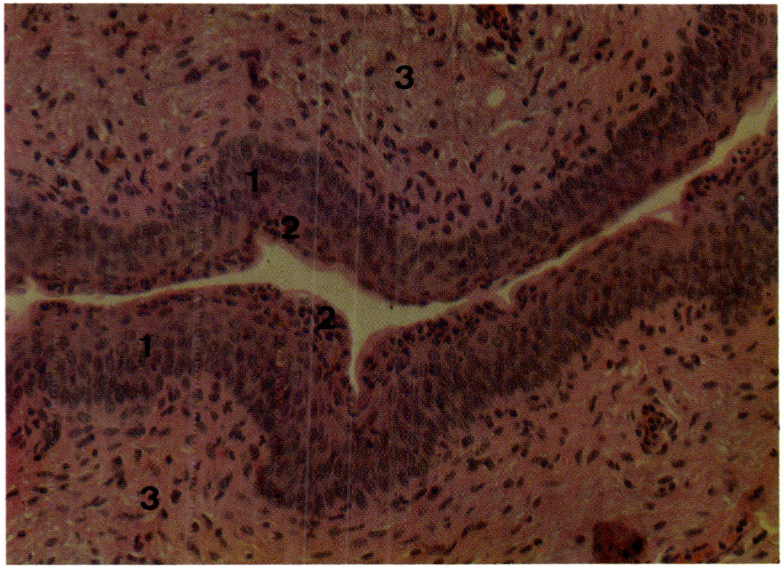

Figure 53

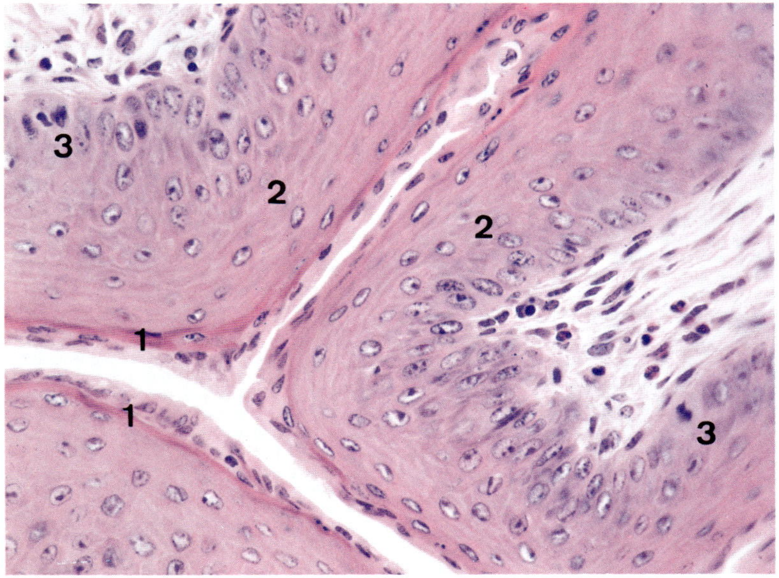

Figure 54

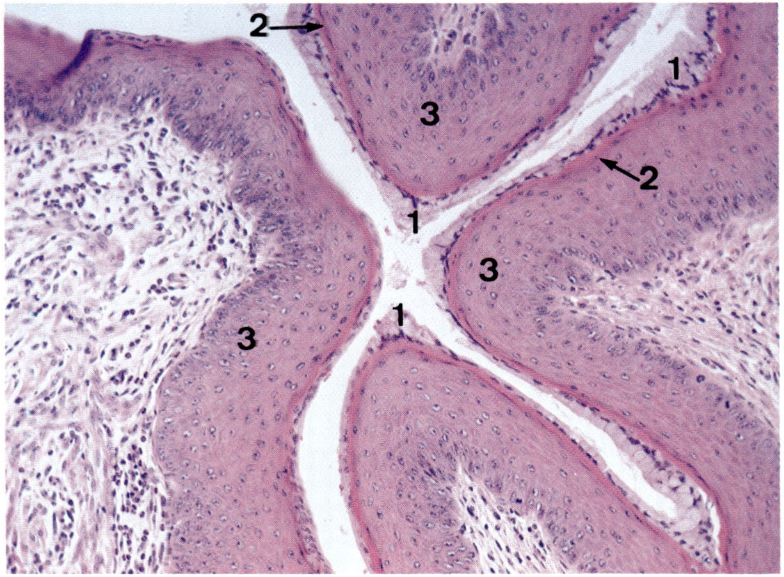

Figure 55

56. Vagina: Proestrus/estrus PAS 250×

Same field as previous H & E section. Mucous cells stimulated by progesterone appear PAS-positive (red).

1. Mucous cells (red)
2. Stratified squamous epithelium

57. Vagina: Estrus H & E 250×

During estrus the cornified granulosum layer thickens and appears superficially. No leukocytes or mitotic figures are seen. Many dead cells and cellular debris are sloughed into the lumen.

1. Cornified layer
2. Cells and cellular debris in lumen
3. Stratified squamous epithelium

58. Vagina: Metestrus I H & E 250×

During metestrus I cornified cells appear in the lumen, with polymorphonuclear leukocytes appearing in the epithelium. The cornified layer desquamates, and many cornified cells appear in the lumen. Leukocytes are present in the epithelium.

1. Cornified layer desquamating into lumen
2. Polymorphonuclear leukocytes
3. Stratified squamous epithelium
4. Smooth muscle and connective tissue

59. Vagina: Metestrus II H & E 250×

An abundance of polymorphonuclear cells are seen in the squamous epithelium before they move out into the lumen with other cellular debris.

1. Polymorphonuclear leukocytes and cornified cells in lumen
2. Leukocytes in epithelium
3. Stratified squamous epithelium

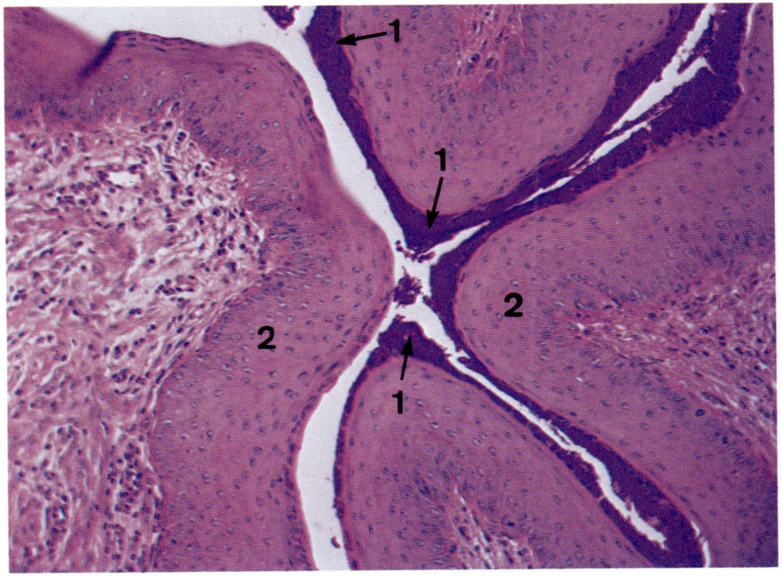

Figure 56

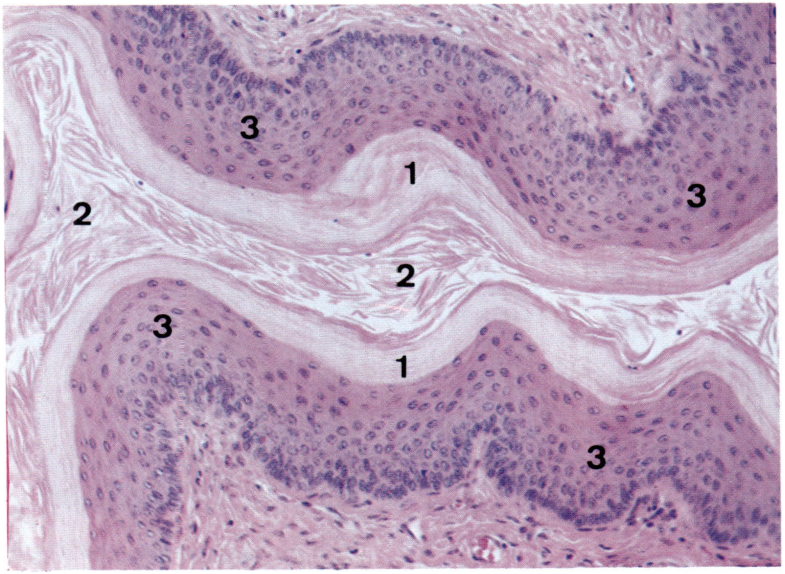

Figure 57

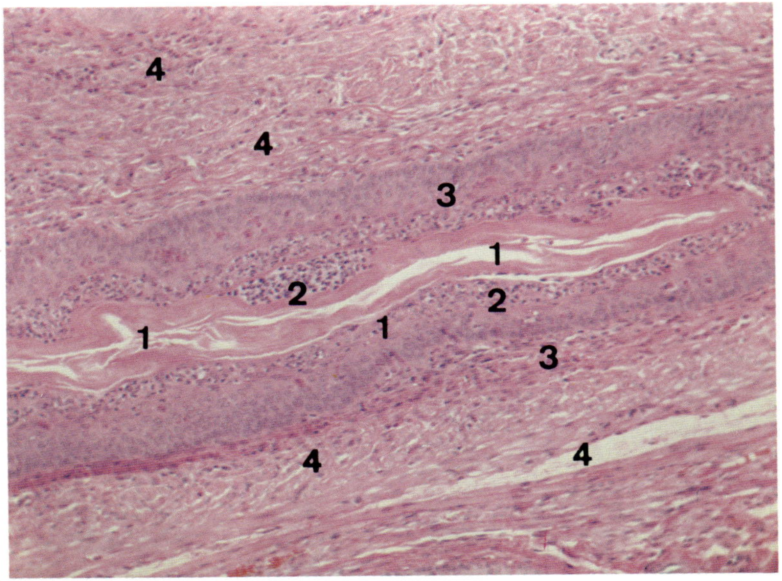

Figure 58

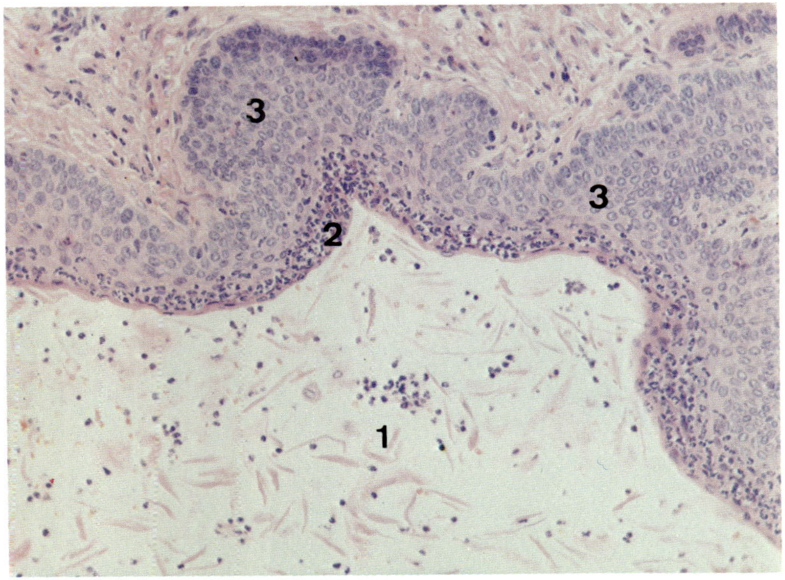

Figure 59

60. Nipple: Lactating breast Heidenhain's Azan 150×

The outer skin surrounds the nipple on the left. The large, oval space center right is the lactiferous duct, below which are lobules. Subcutaneous fat occurs between collagen fibers.

1. Skin
2. Lactiferous duct
3. Lobules
4. Hair follicles
5. Collagen fibers (blue)
6. Sebaceous gland
7. Fat cells

61. Mammary gland: Nonlactating H & E 50×

Mammary tissue in the mouse is found on both sides of the ventral abdominal surface extending from the neck to the groin. This section was made through the outer skin layer through fat and striated muscle to the thick fatty tissue, in which lobes of mammary tissue and intralobular ducts are located. A large vein is at lower right.

1. Epidermis
2. Dermis
3. Hair follicle
4. Fat cells
5. Striated muscle
6. Intralobular ducts
7. Blood-filled vein

62. Mammary gland: Lactating H & E 100×

Thin skin with hair follicles forms the upper border, beneath which are a layer of fat cells and a few strands of skeletal muscle. Most of the gland consists of lobules of secreting alveoli, with two lactiferous ducts in the center.

1. Skin
2. Fat cells
3. Lactiferous duct
4. Lobules of alveoli
5. Striated muscle

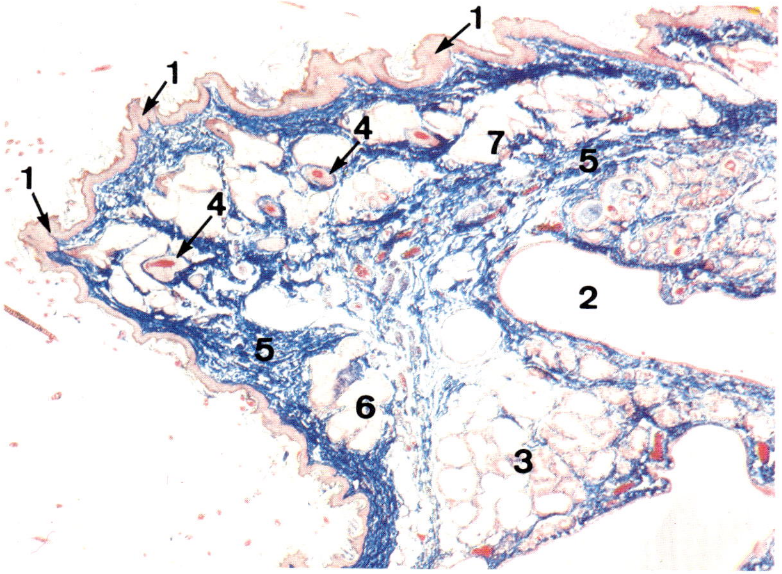

Figure 60

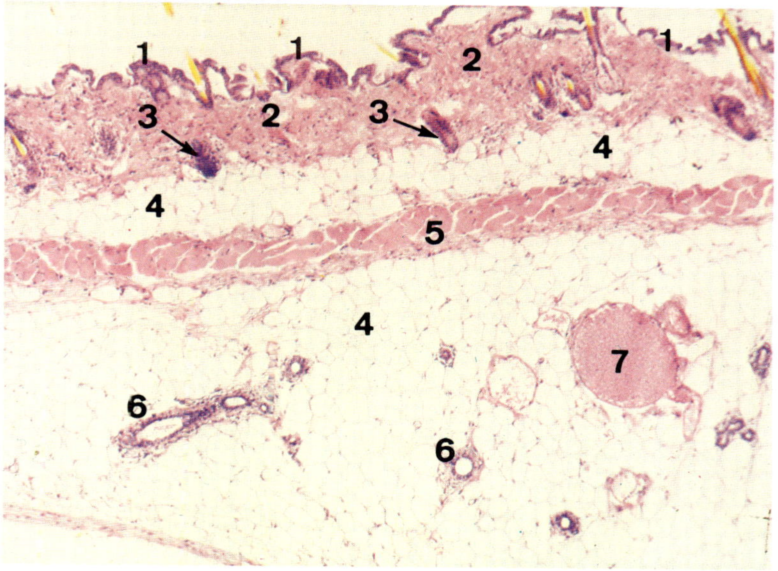

Figure 61

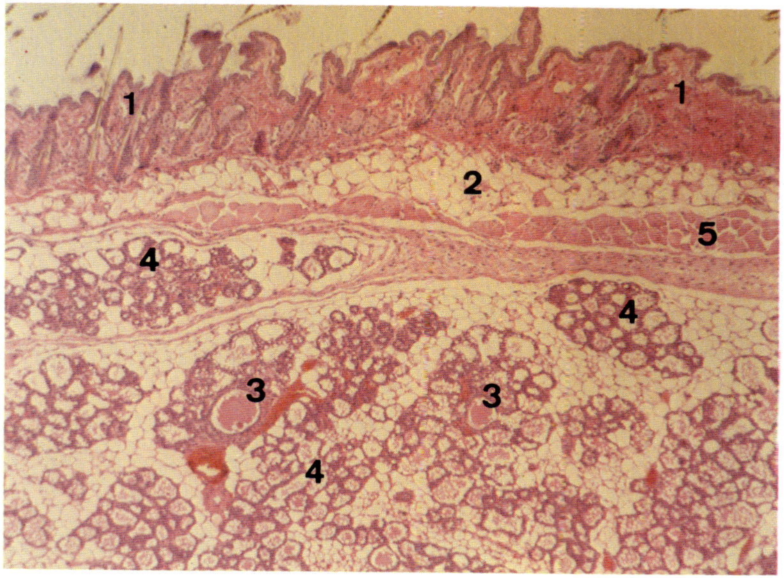

Figure 62

A. Male Urogenital System

1. Left kidney
2. Ureter
3. Left testis
4. Epididymis caput
5. Epididymis cauda
6. Vas deferens
7. Ampullary gland
8. Vesicular gland (seminal vesicle)
9. Coagulating gland
10. Urinary bladder
11. Ventral prostate
12. Dorsal prostate
13. Ischiocavernosus muscle
14. Bulbourethral gland
15. Bulbocavernosus muscle
16. Preputial gland
17. Penis (within sheath)

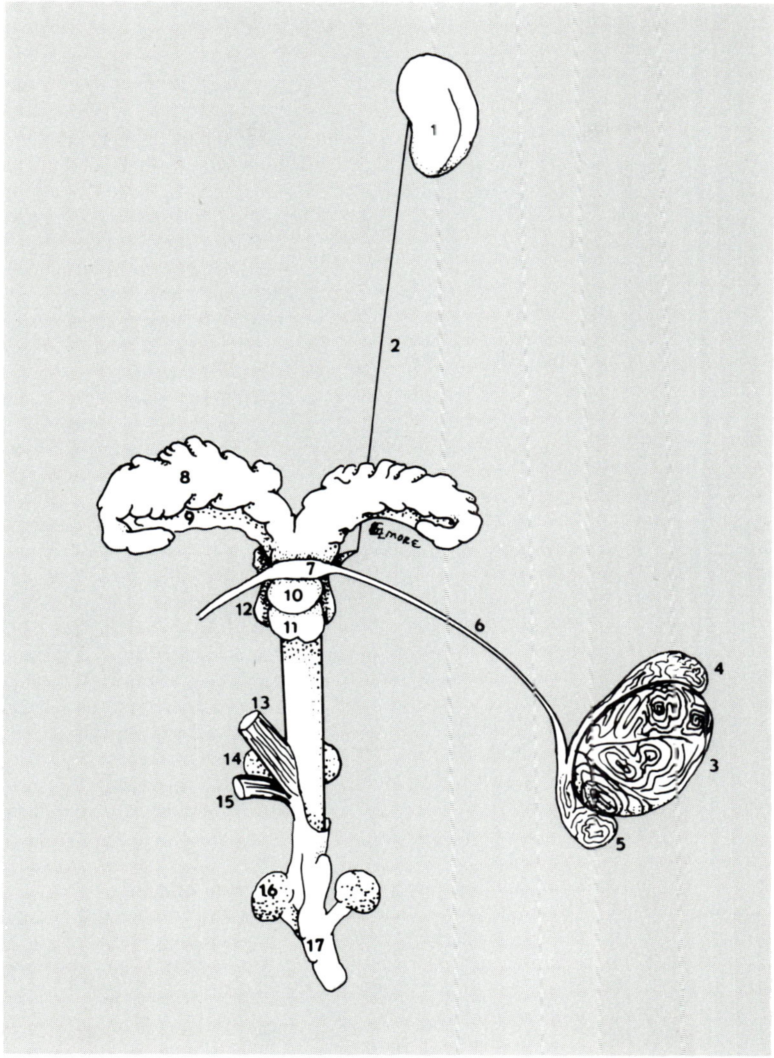

Figure A

63. Penis H & E 50×

In the center of the section is the os penis, below which is the urethra. Surrounding these structures are the corpora cavernosa of connective tissue. Epithelium of the glans completely surrounds the organ, separating it from the preputial cavity. The preputial glands lie on each side of the organ. The skin, with hair follicles and fat cells, forms the outer border.

1. Os penis (bone)
2. Urethra
3. Erectile tissue
4. Epithelium of glans
5. Epithelium of prepuce
6. Preputial cavity
7. Preputial gland
8. Skin

64. Penis Heidenhain's Azan 100×

Starting at the center of this cross section of penis, the os penis lies directly above the urethra, which is lined by transitional epithelium. The vascular erectile tissue (deep blue) surrounding the urethra is the corpora cavernosum. Folded stratified squamous epithelium (brilliant red) forms the prepuce while the entire organ lies in a space lined by specialized epithelium. Preputial glands lie on either side of the penis near the external opening, which is surrounded by the stratified squamous epithelium of the skin. Strands of smooth muscle and fat cells appear within connective tissue.

1. Os penis (bone)
2. Urethral lumen
3. Epithelium of glands
4. Corpora cavernosum
5. Preputial gland
6. Skin
7. Blood vessel
8. Fat cells
9. Epithelium of prepuce

65. Testis H & E 250×

The testis is covered by a thick, fibrous capsule (tunica albuginea). Fibers from the capsule form thick septa. These divide the testis into seminiferous tubules, which lie on a basement membrane containing supporting cells (Sertoli cells) and spermatogenic cells.

1. Seminiferous tubule
2. Interstitial cells of Leydig
3. Developing sperm
4. Capsule—tunica albuginea

66. Testis H & E 250×

This cross section of a seminiferous tubule reveals three types of cells. The supporting, nonproliferating Sertoli cells are located immediately adjacent to the basement membrane and are distributed among spermatogonia. They are the most primitive sperm cells. The next layer of cells is composed of primary spermatocytes, which undergo meiosis and form secondary spermatocytes, located in the third layer of cells. Spermatids are now produced, which form mature spermatozoa. In the interstitial tissue, amid connective tissue and blood vessels, are cells of Leydig, the site of origin of testosterone.

1. Sertoli cell
2. Interstitial cells of Leydig
3. Spermatogonium
4. Primary and secondary spermatocytes
5. Spermatids and mature sperm

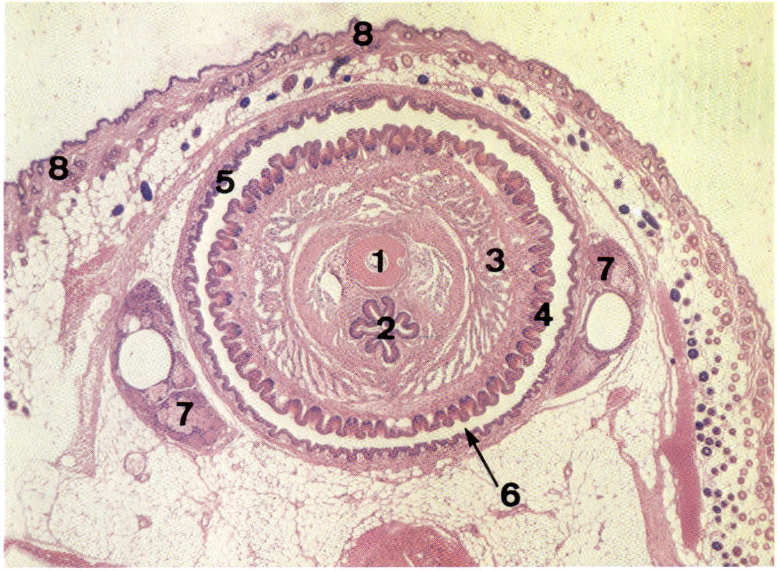

Figure 63

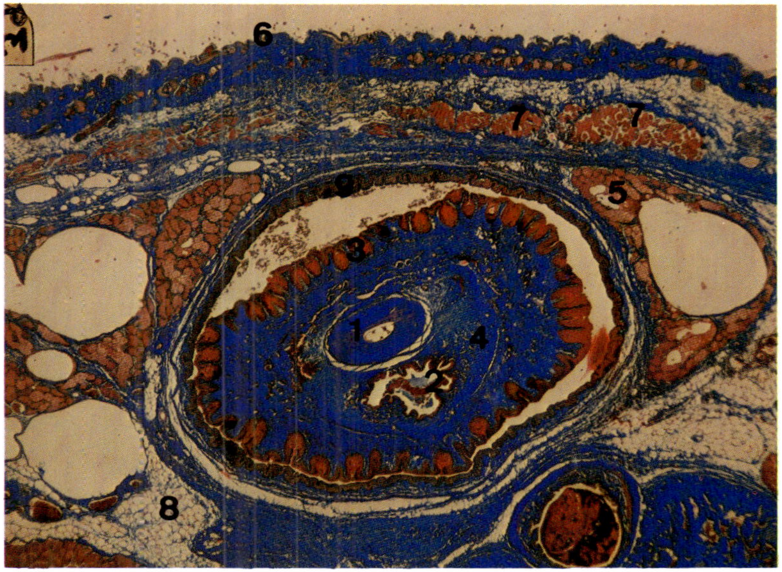

Figure 64

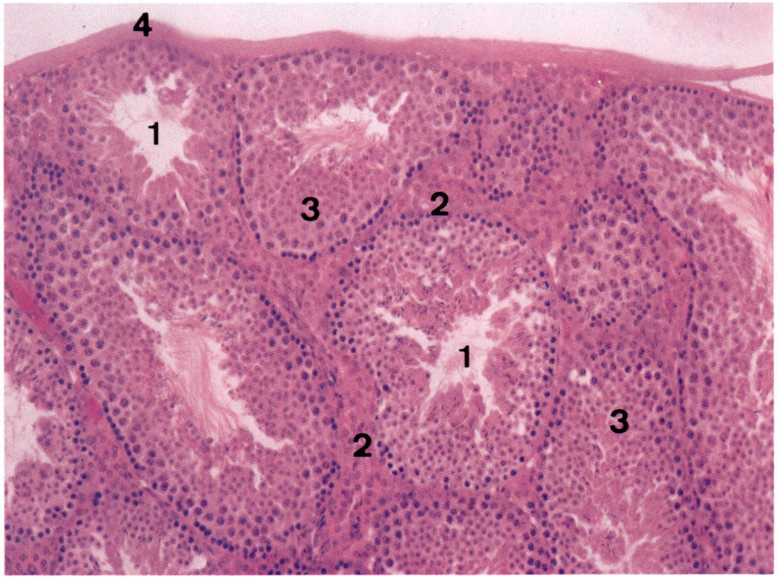

Figure 65

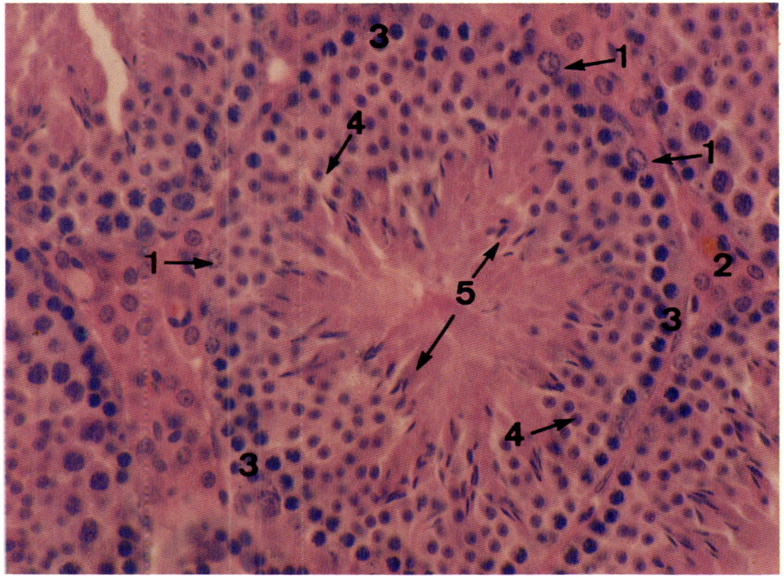

Figure 66

67. Testis — Heidenhain's Azan — 500×

In this section, brilliant red mature spermatozoa occupy the lumen of a seminiferous tubule. Developing stages can be identified from spermatogonia lining the periphery through cells moving toward the lumen, where spermatids are forming mature sperm.

1. Sertoli cell
2. Spermatogonium
3. Primary spermatocyte
4. Secondary spermatocyte
5. Spermatozoa
6. Interstitial cells of Leydig

68. Ductus deferens — H & E — 200×

The duct is filled with spermatozoa and lined by simple columnar epithelium. Smooth muscle and connective tissue surround the ducts.

1. Columnar epithelium
2. Smooth muscle fibers
3. Mass of spermatozoa in lumen

69. Epididymis: Caput — H & E — 250×

The head of the epididymis contains many ducts lined by columnar epithelium with stereocilia on their free surfaces. The lumina of many tubules contain spermatozoa. Fibrous connective tissue separates the tubules.

1. Lumen with spermatozoa
2. Interstitial tissue
3. Capsule

70. Efferent duct — H & E — 500×

Coiled segments are lined by tall columnar epithelium with stereocilia projecting from the free surface of the cells. Several dividing cells, as well as spermatozoa, are present in the lumen.

1. Duct
2. Stereocilia
3. Columnar epithelium
4. Cells in mitosis

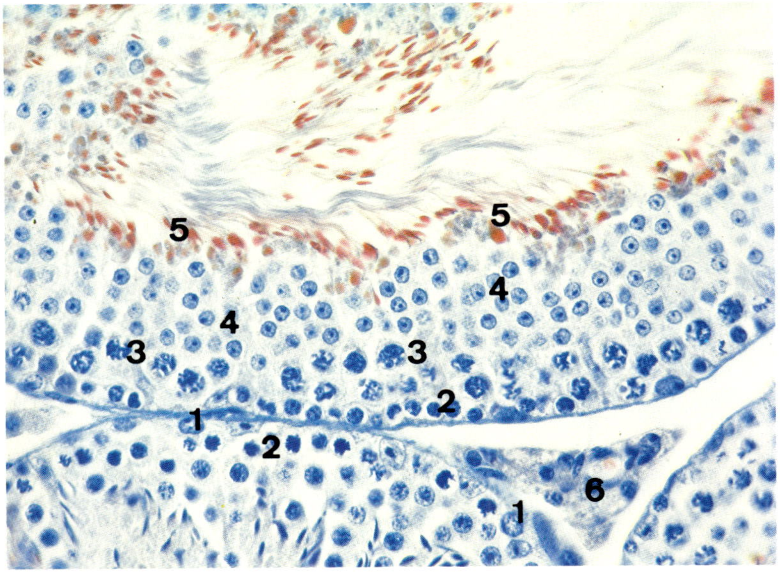

Figure 67

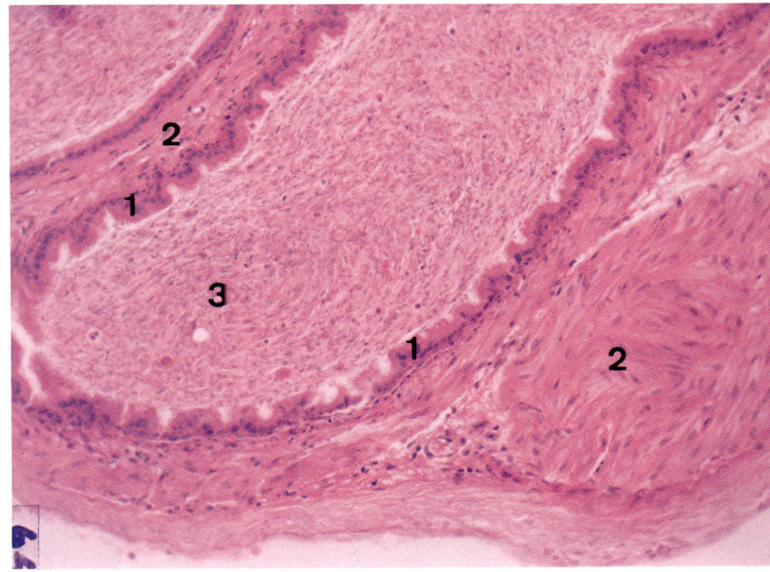

Figure 68

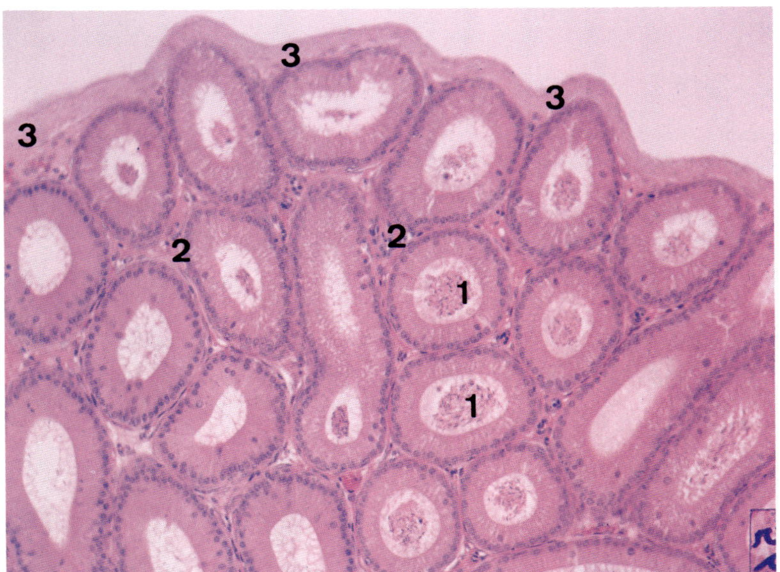

Figure 69

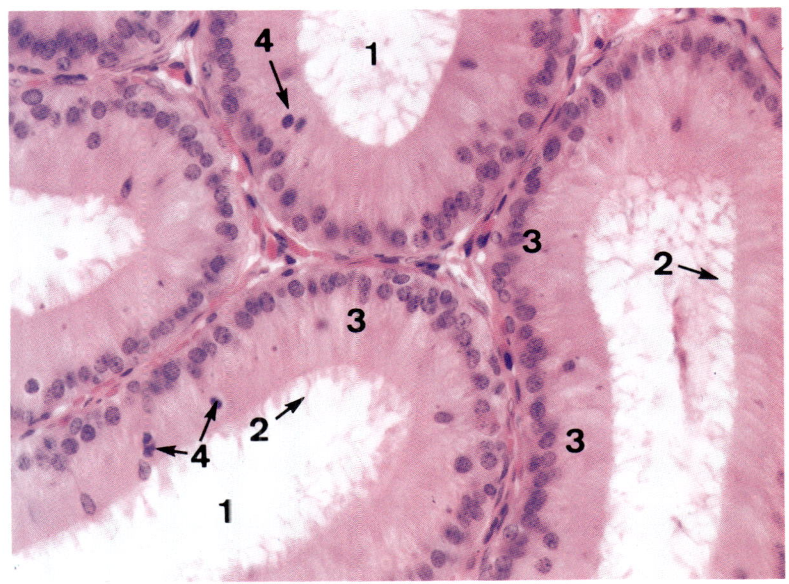

Figure 70

71. Bulbourethral gland H & E 100×

This gland is characterized by alveoli of tall columnar cells, dark nuclei, and very light cytoplasm. A rim of striated muscle partially surrounds the gland.

1. Striated muscle
2. Alveoli—tall columnar cells

72. Male reproductive system H & E 100×

The vas deferens, lined by columnar epithelium and surrounded by smooth muscle, occupies the upper left. Connective tissue with fat cells separates the vas from the ventral prostate in the lower right corner of the field.

1. Vas deferens
2. Smooth muscle and connective tissue
3. Ventral prostate
4. Adipose tissue

73. Male reproductive system H & E 50×

A section of rectum containing a lymphoid nodule lies at the extreme right. Left of the rectum is the urethra, surrounded by smooth muscle and containing pink residue in the lumen. The ventral prostate occupies the left center of the field next to the urethra. At upper left is a cross section of vas deferens; the urinary bladder lies at the lower left.

1. Rectum
2. Urethra
3. Ventral prostate
4. Urinary bladder
5. Pelvic bone
6. Vas deferens

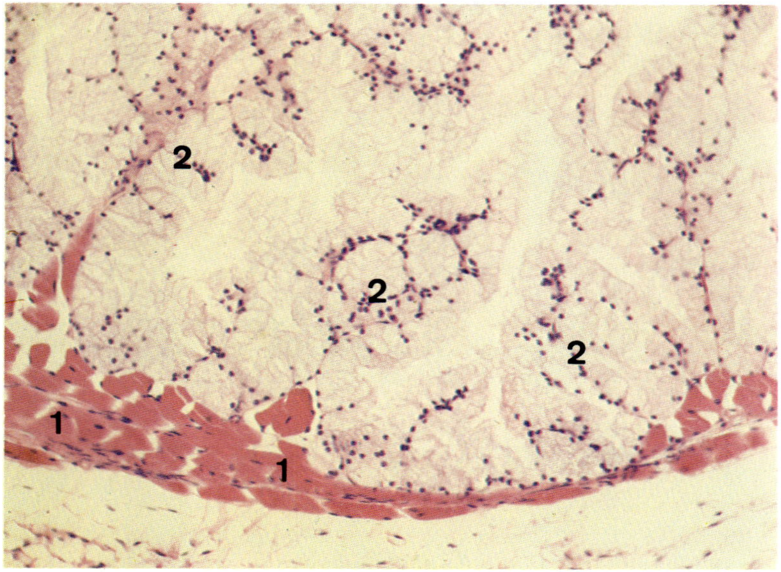

Figure 71

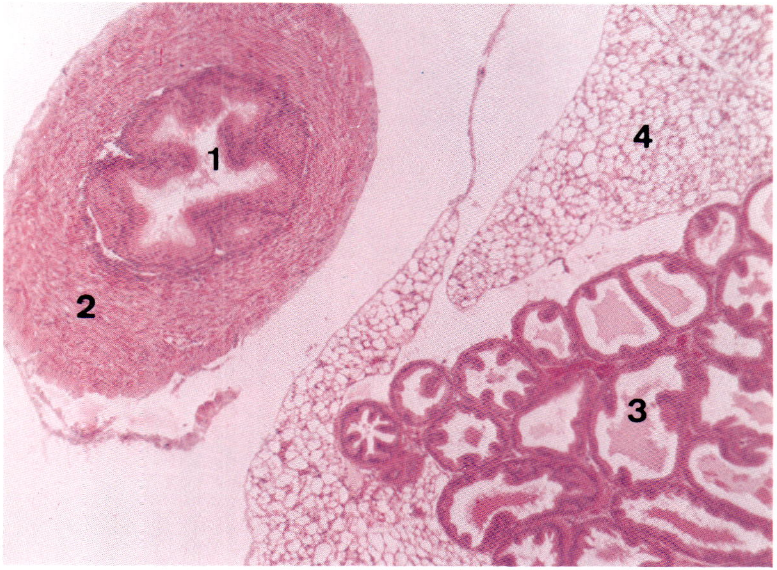

Figure 72

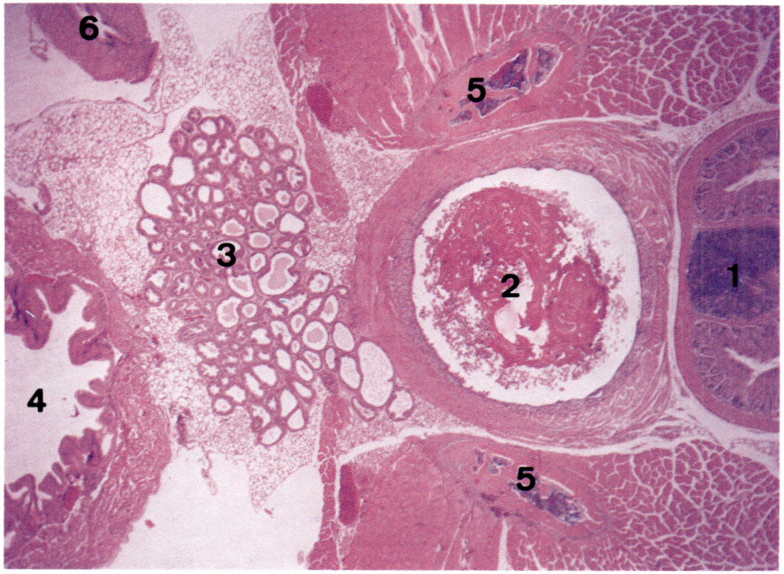

Figure 73

74. Male reproductive system H & E 50×

Red secretion and large alveoli with infolding of mucosa characterize the vesicular gland; the coagulating gland immediately adjacent is filled with pink secretory material and lightly stained glandular tissue.

1. Vesicular gland
2. Coagulating gland

75. Male reproductive system H & E 100×

At a higher magnification the differences in epithelial linings of the coagulating gland (above) and vesicular gland (below) are clearly seen. The characteristic red secretion of the vesicular gland is apparent.

1. Vesicular gland
2. Coagulating gland

76. Male reproductive system Heidenhain's Azan 500×

Columnar epithelium and red-stained secretion of the vesicular gland may be compared with cuboidal epithelium and pink secretion of the coagulating gland.

1. Vesicular gland
2. Coagulating gland

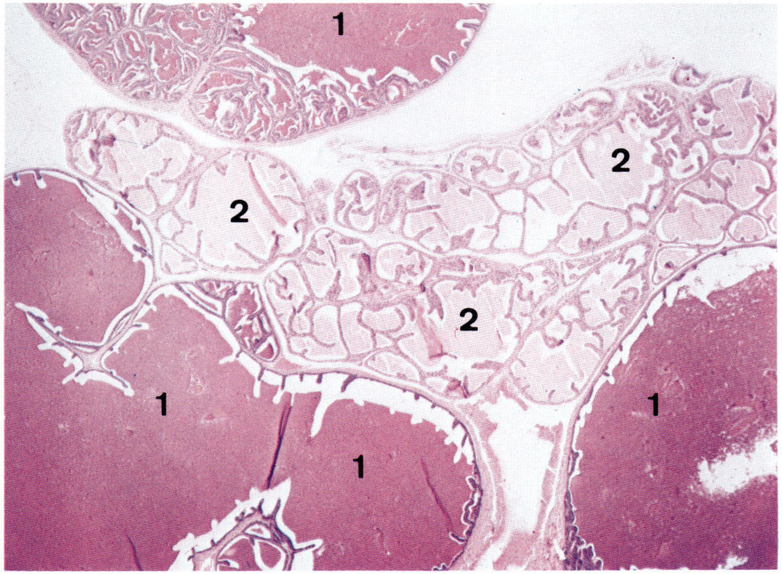

Figure 74

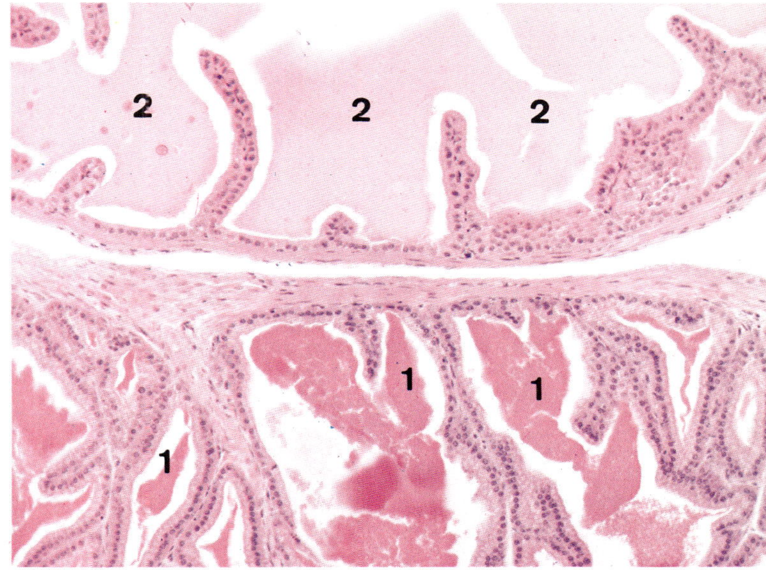

Figure 75

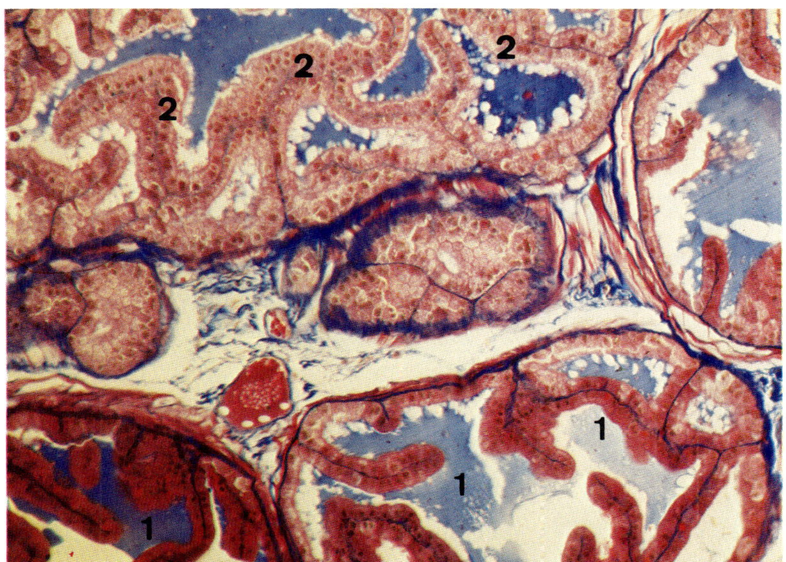

Figure 76

77. Tongue: Posterior dorsal surface — H & E — 100×

The surface of the tongue in this region is covered with stratified squamous epithelium. Two taste buds are present in the center. Above the epithelium are bundles of striated muscle fibers. At upper left and in the center of the muscle are two groups of serous glands; mucous glands are seen at right.

1. Mucous glands
2. Serous glands
3. Stratified squamous epithelium
4. Taste buds
5. Striated voluntary muscle

78. Taste bud — H & E — 500×

A taste bud on the dorsal surface of the tongue in which the principal internal structures are easily seen.

1. Stratified squamous epithelium
2. Gustatory cell
3. Sustentacular cell
4. Location of pore

79. Filiform papillae: Tongue — Masson stain — 500×

Pointed filiform papillae extend from a modified stratified squamous epithelium. The other four layers (comparable to thick skin) are easily identified. Green collagen fibers and cross sections of striated muscle are immediately adjacent to and below the basal layer. A dermal papilla projects up into the epithelium on the left.

1. Filiform papilla
2. Stratum granulosum
3. Stratum germinativum
4. Basal layer of germinativum
5. Dermal papilla—collagen fibers (green)
6. Striated muscle
7. Nerve trunk

80. Taste buds — Heidenhain's Azan — 500×

A group of taste buds in the tongue illustrate internal structures. The pores empty into a lumen exiting on the dorsal surface of the tongue.

1. Stratified squamous epithelium
2. Taste bud pore
3. Taste bud
4. Collagen fibers (blue)

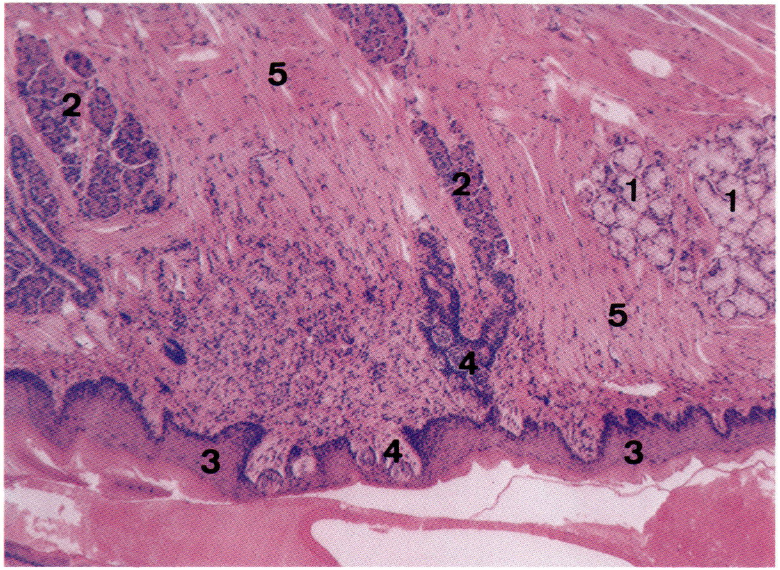

Figure 77

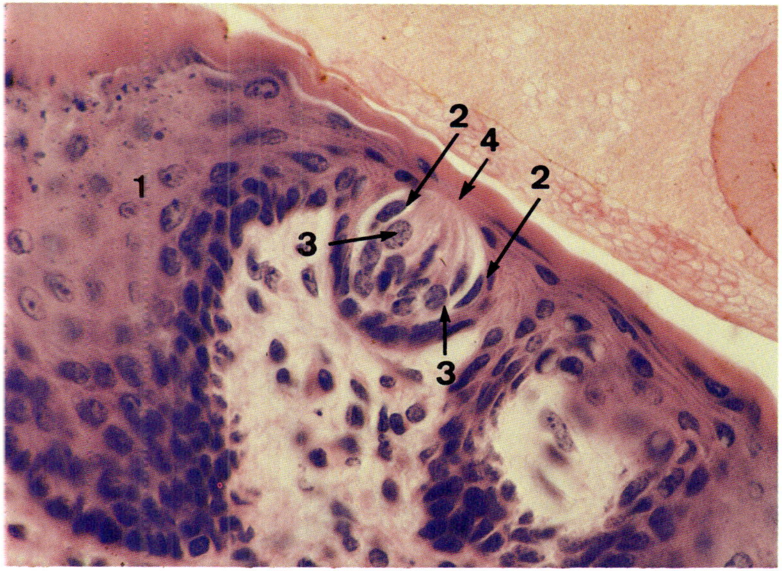

Figure 78

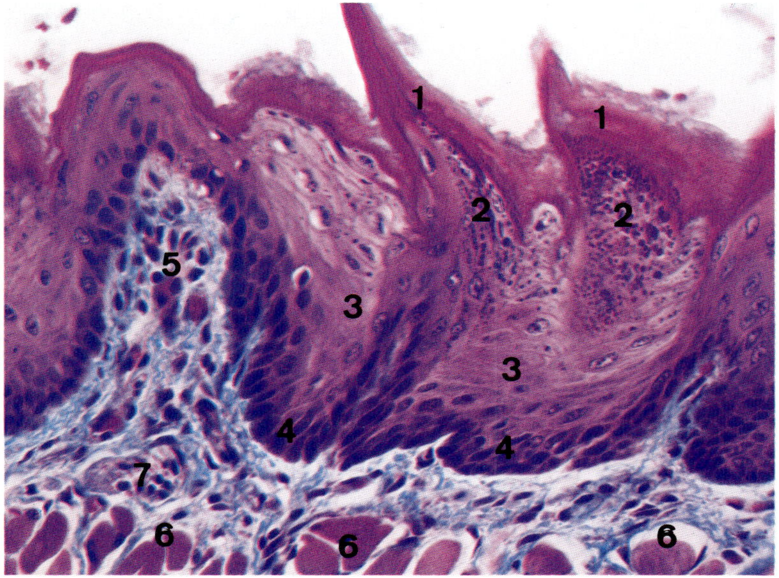

Figure 79

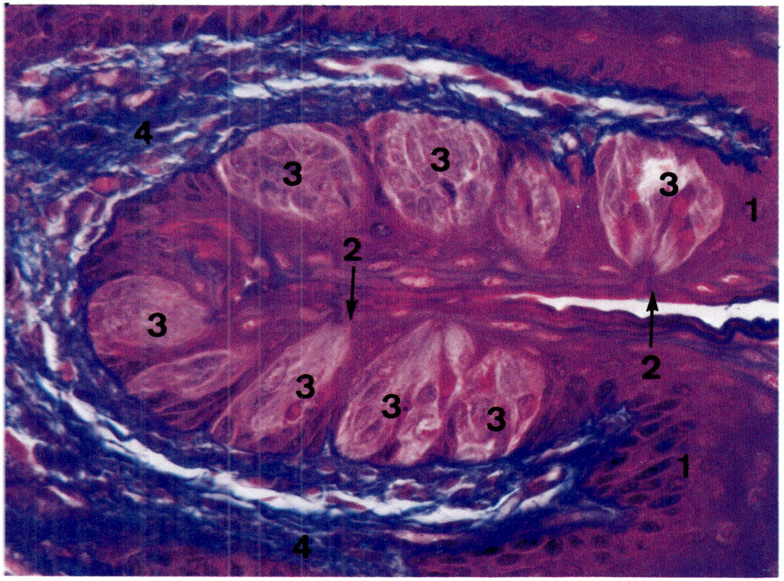

Figure 80

81. Esophagus H & E 100×

In this section the esophagus consists of folds surrounding a space, the lumen. The epithelial lining is stratified squamous; surrounding it is subepithelial connective tissue. Striated muscle completely surrounds the entire structure.

1. Stratified squamous epithelium
2. Lamina propria—connective tissue
3. Striated muscle

82. Esophagus H & E 500×

Several layers of keratinized, stratified squamous epithelium can be identified. One cell in the basal layer is undergoing division. Striated muscle fibers in a longitudinal plane mixed with connective tissue and blood vessels occupy the area below the epithelium.

1. Stratified squamous epithelium
2. Dividing cell in basal layer of epithelium
3. Striated muscle
4. Lamina propria—connective tissue

83. Salivary glands H & E 50×
Lymph node

Three types of salivary glands can be identified. Immediately adjacent to and above the lymph node in the upper center is the serous parotid gland. The submandibular gland to the left of the parotid is of the mixed type, serous and mucous. The mucus-secreting sublingual gland is the lightly stained gland at center right.

1. Parotid gland (serous)
2. Lymphatic nodule
3. Lymphatic vessels
4. Sublingual gland (mucous)
5. Submandibular gland (mixed)
6. Blood vessel

84. Salivary glands PAS 150×

The three types of salivary glands are easily differentiated: sublingual, center to left (mucus-secreting) (brilliant red); parotid (serous), lower right; and submandibular (mixed serous and mucous), upper left and center.

1. Mucous type (sublingual gland)
2. Serous type (parotid gland)
3. Mixed type (submandibular)

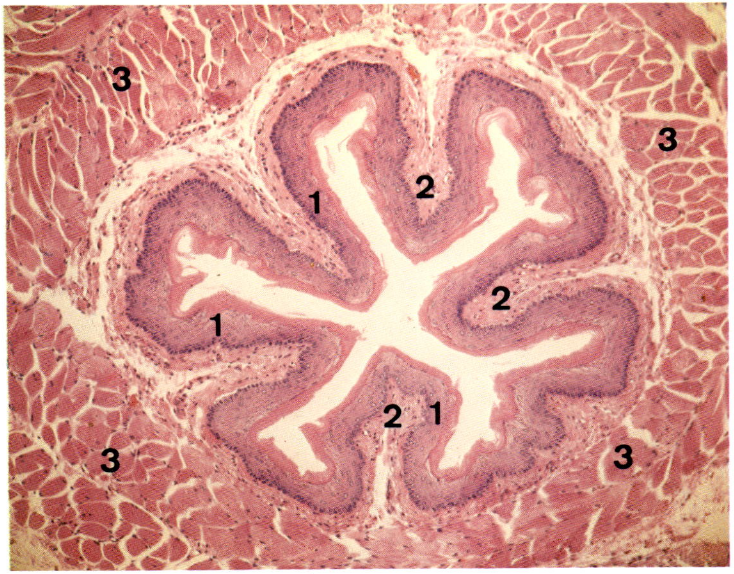

Figure 81

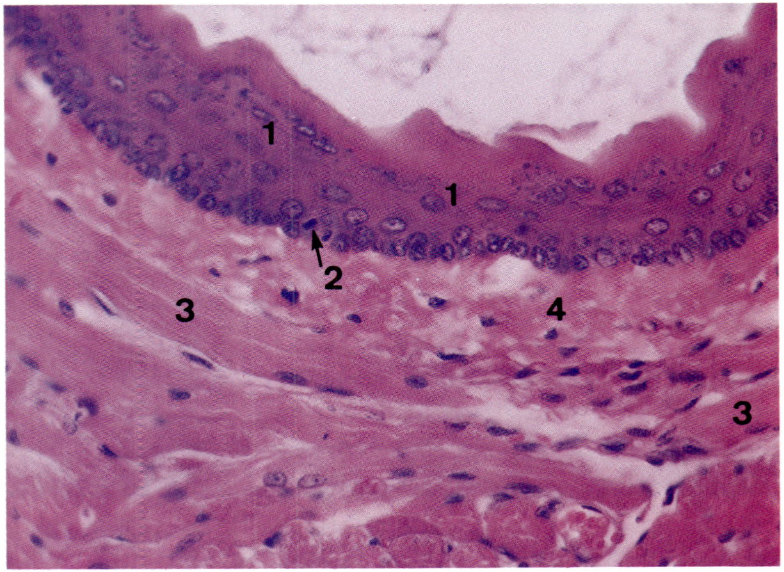

Figure 82

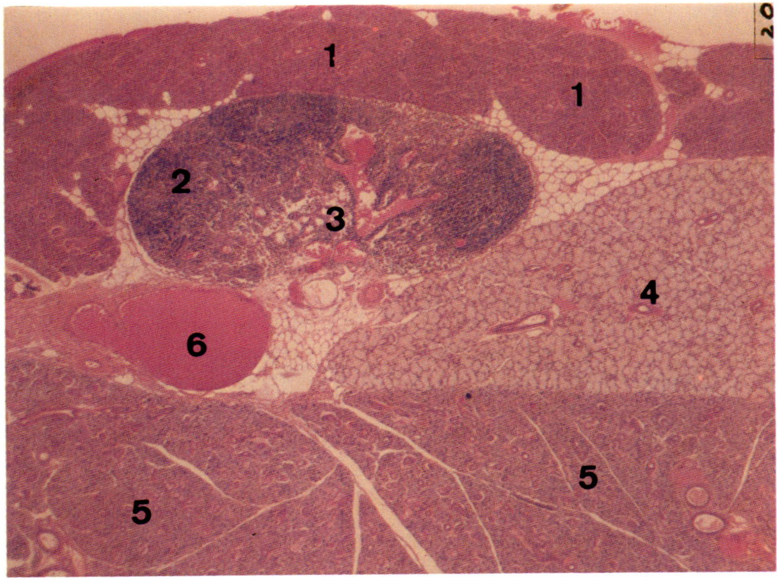

Figure 83

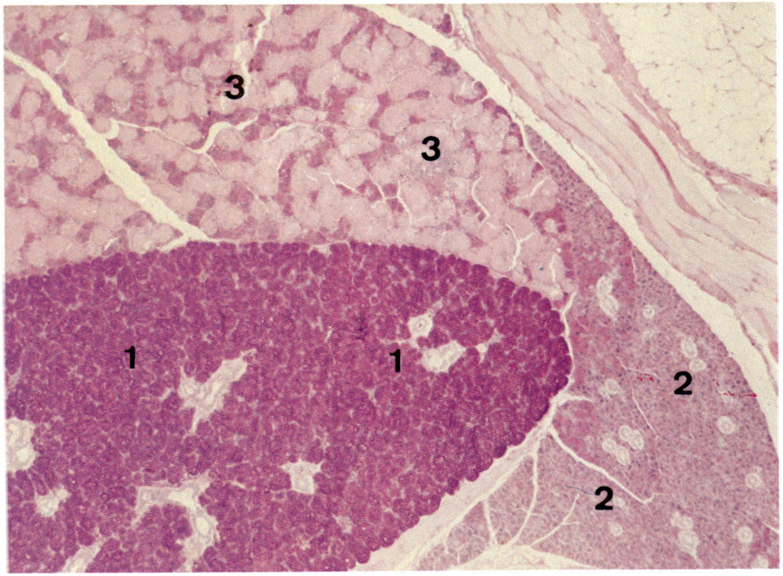

Figure 84

85. Submandibular gland, male — Methylene Blue, Basic fuchsin — 500×

Histological structures characteristic of the adult male mouse are apparent in this 1.5-μm-thick methacrylate section. The nuclei of the columnar intercalated tubular cells are located at the base of the cell. Bright red secretory granules fill the cytoplasmic area between the nucleus and lumen of the duct.

1. Basally located nucleus
2. Secretory granules
3. Alveolus

86. Submandibular gland, female — Methylene Blue, Basic fuchsin — 500×

In adult females and young males the nuclei of tubular cells are located in the center of the cell. Secretory granules are absent or are very sparse in the cytoplasm of this methacrylate section.

1. Centrally located nucleus
2. Alveolus
3. Pink-staining mitochondria

87. Stomach, glandular and nonglandular — H & E — 100×

Keratinized stratified squamous epithelium forms the nonglandular portion of the stomach. There is an abrupt change to a modified columnar epithelium, forming the glandular portion. Submucosa of loose connective tissue is seen between the mucosa and circular smooth muscle layer, next to which is the layer of longitudinal muscle. The thin outer covering is serosa.

1. Stratified squamous epithelium (keratinized)
2. Glands
3. Smooth muscle—circular
4. Smooth muscle—longitudinal
5. Submucosa—connective tissue
6. Serosa

88. Stomach glands — Alcian Blue, Tartrazine — 250×

In the glandular portion of the stomach three types of cells can be identified by special stains. The submucosa lies immediately beneath the glandular mucosa. At the base of the mucosa are chief or zymogenic cells, stained brown. The spherical cells with small round nuclei and blue cytoplasm occupying most of the middle portion are parietal cells. The cells located near the top, colored green, are mucous neck cells.

1. Chief cell layer (dark brown)
2. Parietal cell layer (blue) (tan spherical central nucleus)
3. Mucous neck cells (light green)

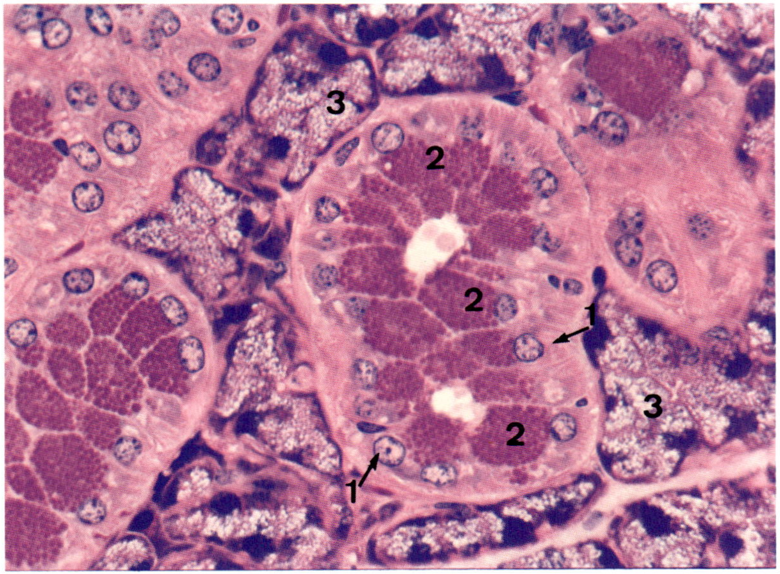

Figure 85

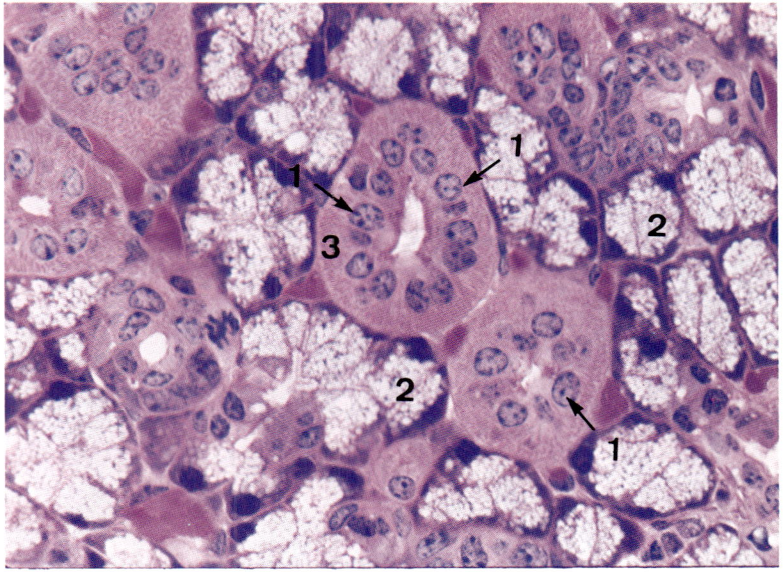

Figure 86

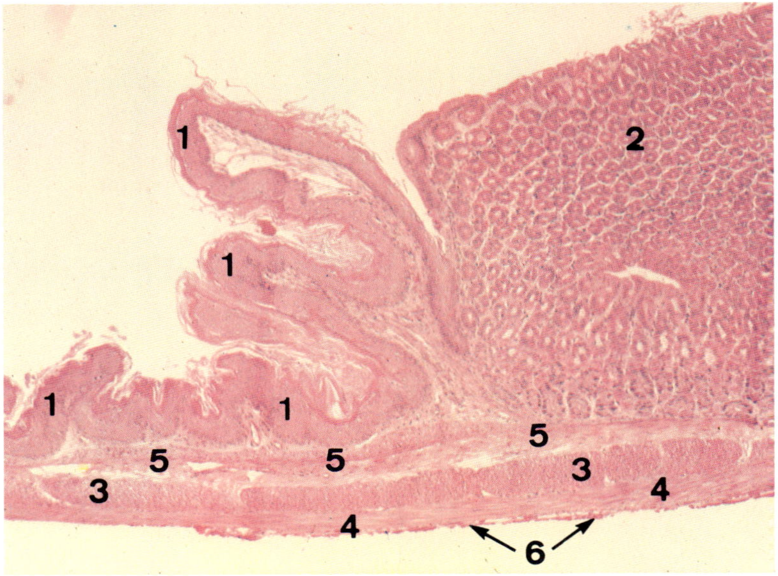

Figure 87

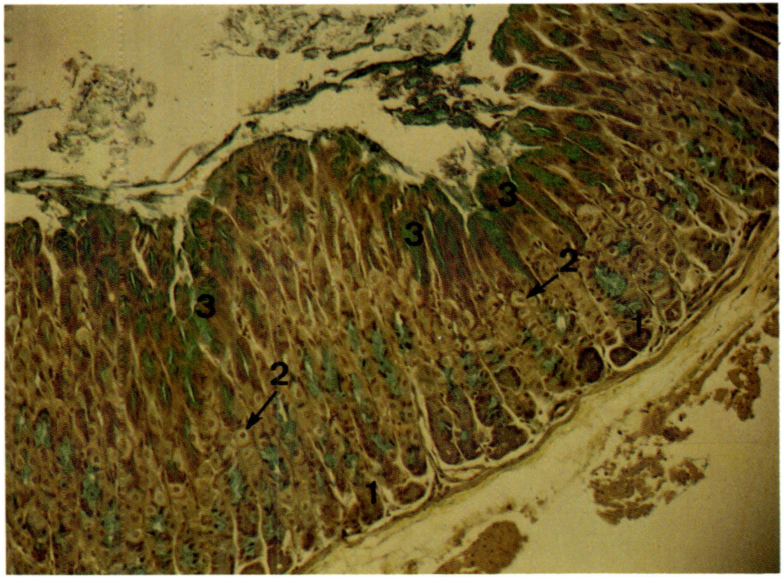

Figure 88

89. Stomach mucosa H & E 500×

Spherical cells with round, centrally located nuclei and pink cytoplasm are parietal cells. Chief or zymogenic cells are grouped near the bottom of the mucosa and contain dark blue nuclei and gray cytoplasm.

1. Chief cells
2. Parietal cells
3. Submucosa

90. Stomach: Duodenum Hematoxylin–PAS 75×

The glandular stomach with PAS-positive mucoid secretion occupies the left half and center of the section. At the junction of the pyloric end of the stomach with the duodenum, the lumen widens as the epithelium becomes simple columnar. Brunner's glands in the submucosa of the duodenum react positively (red) to periodic acid–Schiff reagent.

1. Smooth muscle—stomach wall
2. Stomach glands, mucous (red)
3. Pyloric opening
4. Brunner's glands of duodenum
5. Villi—duodenum
6. Pancreas

91. Stomach: Duodenum Hematoxylin–PAS 200×

At higher magnification the structural differences between mucous glands of the stomach and Brunner's glands and columnar epithelium of the duodenum become apparent.

1. Stomach glands, mucous (red)
2. Villi—duodenum
3. Brunner's glands—duodenum
4. Fat cells
5. Smooth muscle—stomach wall

92. Small intestine: Ileum H & E 150×

Peyer's patch is a collection of lymphocytes grouped into a lymphoid nodule. Aggregates of lightly stained cells occupy germinal centers, from which mature lymphocytes proliferate. Cross sections of villi of varying lengths occupy the upper portion of the region.

1. Lymphoid nodule (Peyer's patch)
2. Germinal center
3. Serosa
4. Villi
5. Lumen with fibrous debris

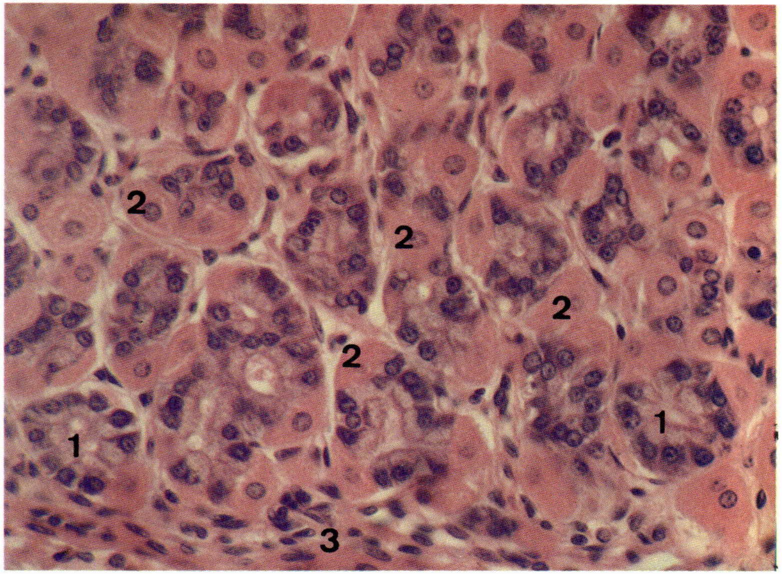

Figure 89

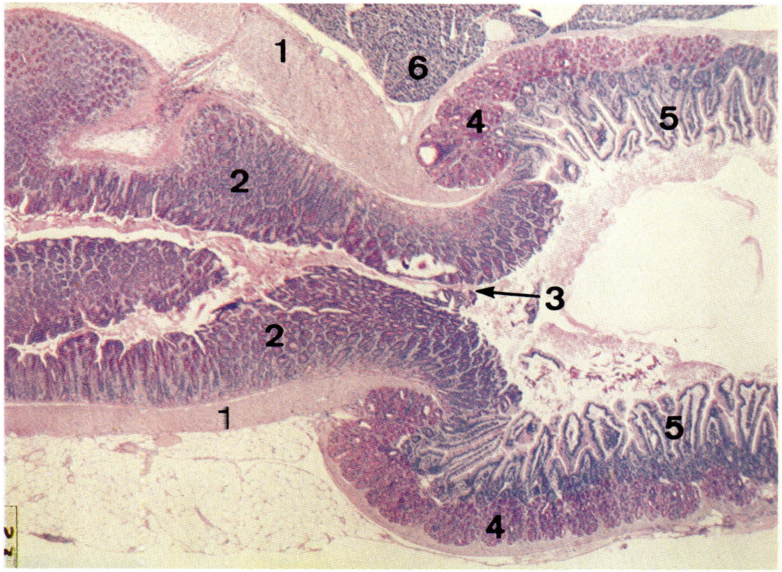

Figure 90

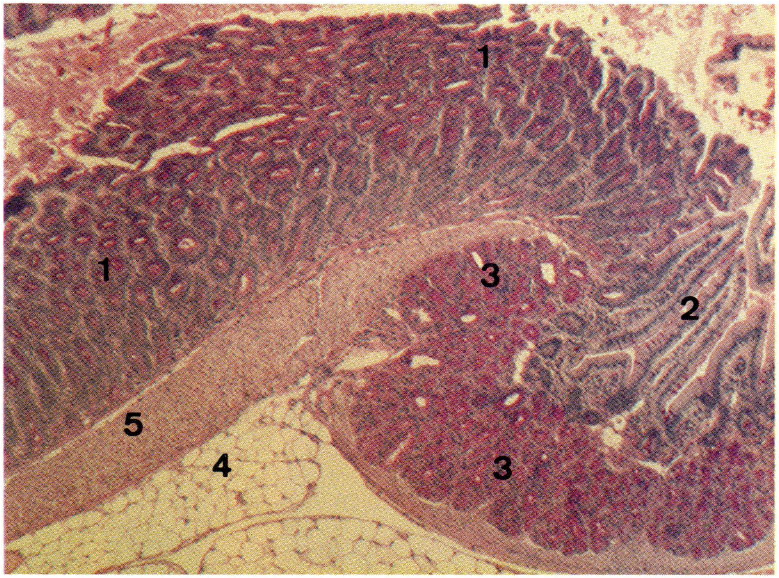

Figure 91

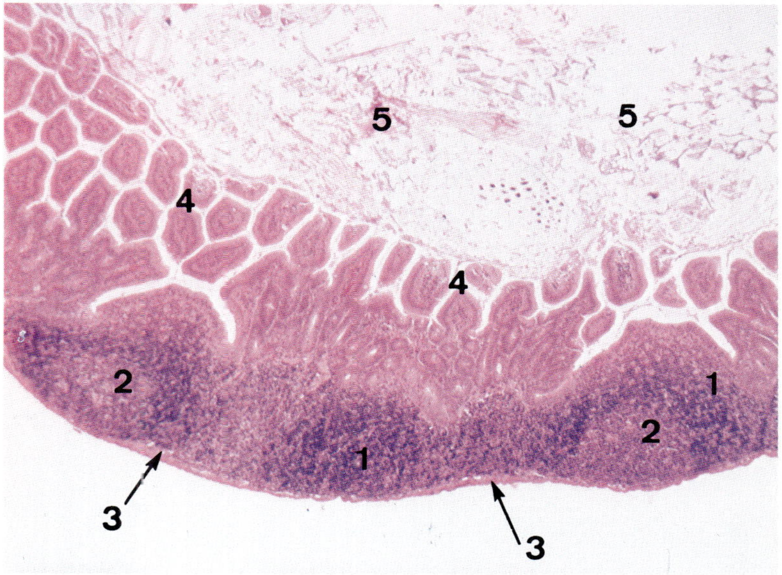

Figure 92

93. Small intestine: Ileum — H & E — 500×

Lymphocytes forming part of Peyer's patch appear in the submucosa below the connective tissue. Extending to the left are four crypts; at the base of each are brick-red granules of Paneth cells.

1. Paneth cell granules (orange-red)
2. Submucosa
3. Crypt of Lieberkühn
4. Goblet cell
5. Lymphoid nodule (Peyer's patch)

94. Villi: Jejunum — Methacrylate, Methylene Blue, Basic fuchsin — 100×

This low-power view shows the relationship between the fingerlike villi and crypts of Lieberkühn. Cross sections of crypts are seen on the right adjacent to smooth muscle.

1. Villus
2. Lamina propria
3. Crypt of Lieberkühn
4. Goblet cell
5. Smooth muscle

95. Villus — Methacrylate, Methylene Blue, Basic fuchsin — 500×

In this 1.5-μm thin section, internal structures of the villus are apparent. A prominent striated border outlines the lumen side of the simple columnar epithelium. Mitochondria, darkly stained, occupy the distal portion of each epithelial cell. Goblet cells appear as spaces in the epithelium.

1. Simple columnar epithelium
2. Mitochondria
3. Striated border
4. Goblet cell
5. Lamina propria
6. Blood vessel

96. Cecum — H & E — 60×

The extension of intestine from the ileum to the beginning of the large intestine is the cecum. The mucosa, occurring in many folds, is lined by columnar epithelium. Beneath the submucosa is a layer of smooth muscle covered by serosa.

1. Mucosa
2. Smooth muscle
3. Intestinal contents

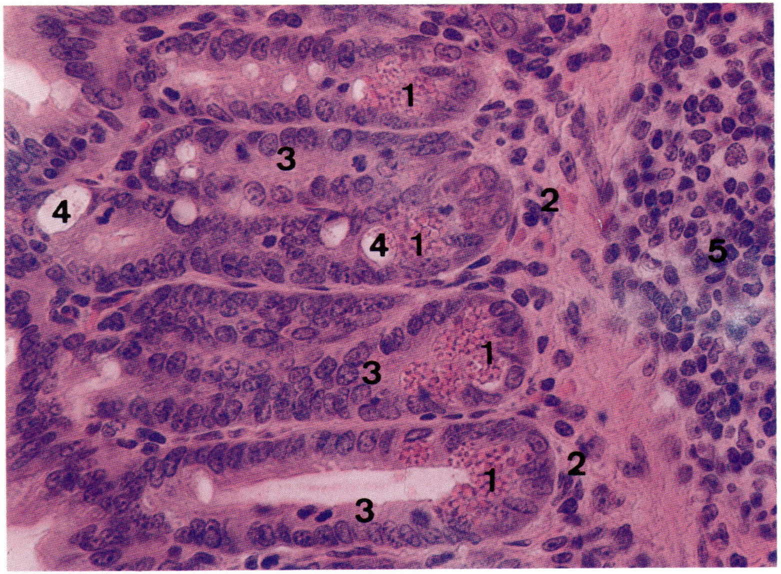

Figure 93

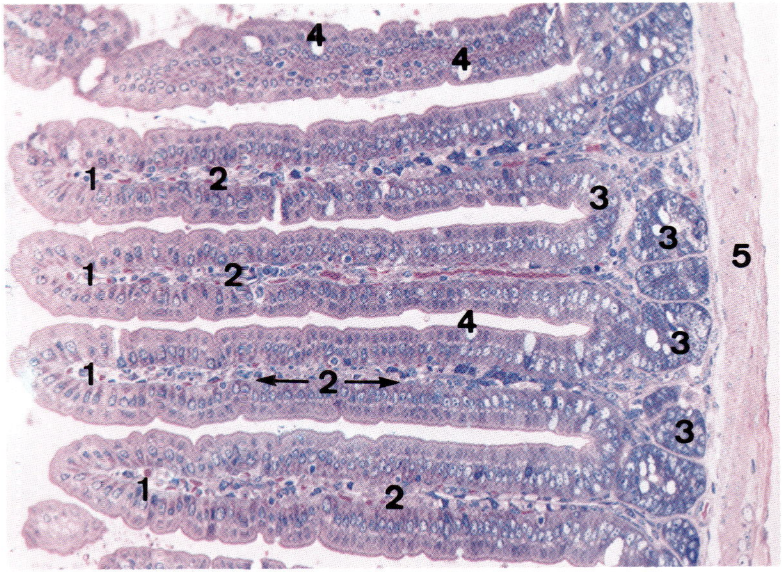

Figure 94

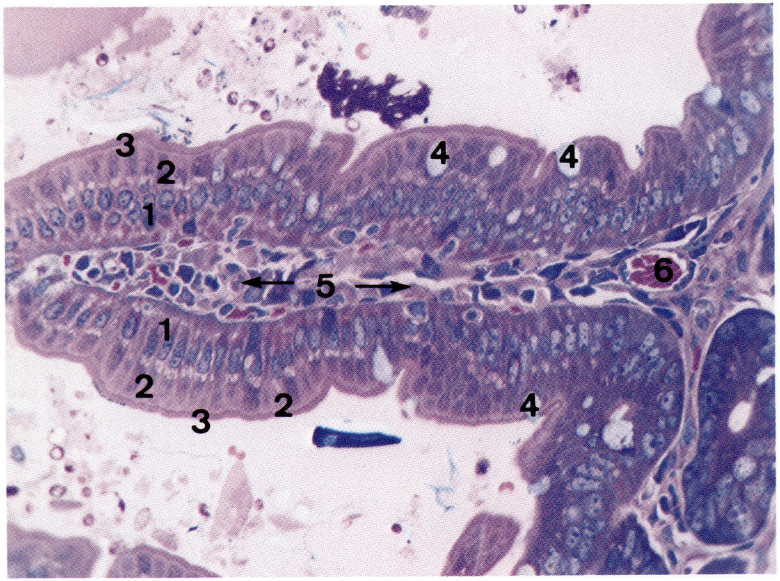

Figure 95

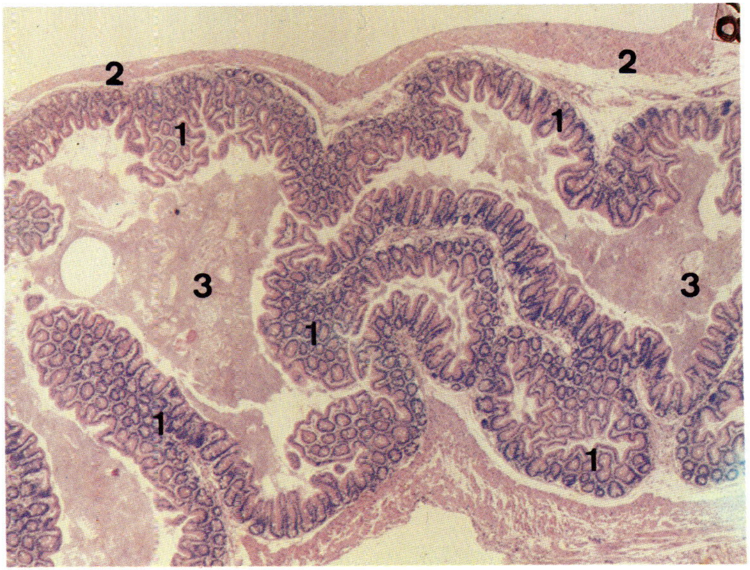

Figure 96

97. Large intestine PAS 500×

Many mucus-filled goblet cells appear bright red. The absence of villi and Paneth cells also aids in identifying the large intestine. Layers of smooth circular and longitudinal muscle are covered by the outermost thin layer, the serosa.

1. Mucus in epithelial cells (red)
2. Submucosa
3. Smooth muscle
4. Serosa

98. Anus H & E 50×

The anal opening is bounded by stratified squamous epithelium of the skin. Columnar epithelium of the rectum abruptly changes to stratified squamous. Striated muscle forms the anal sphincter.

1. Skin
2. Stratified squamous epithelium
3. Simple columnar epithelium
4. Striated muscle
5. Sebaceous glands

99. Pancreas H & E 250×

Surrounded by dark red acinar cells, three lightly stained islets of Langerhans appear at left, right, and lower center. An epithelial-lined pancreatic duct is adjacent to the left islet, and portions of an artery and vein appear at upper left and right.

1. Pancreatic duct
2. Artery (with vein at upper right)
3. Islet of Langerhans
4. Acinar cells

100. Pancreas Modified aldehyde–fuchsin 250×

In this section the alpha and beta cells of the islets of Langerhans are differentiated: beta cells are purple and alpha cells are red.

1. Islets of Langerhans
2. Purple beta cells
3. Red alpha cells
4. Acinar cells
5. Pancreatic duct
6. Blood vessel

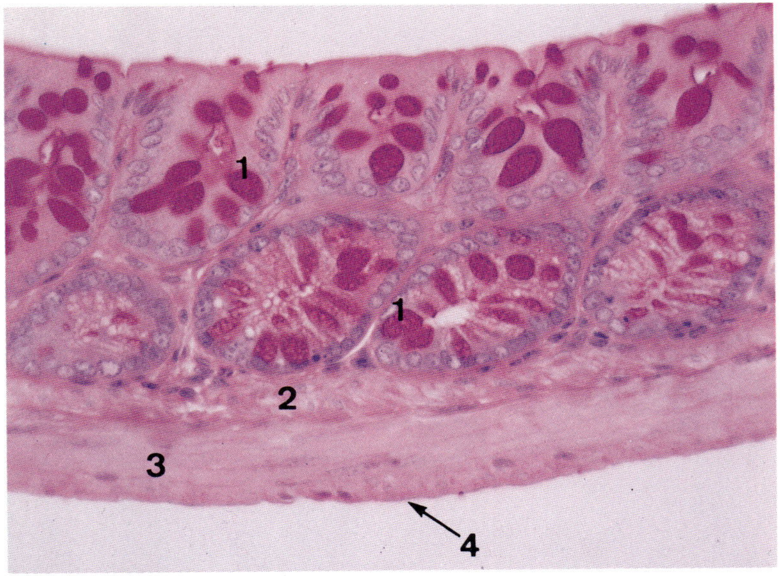

Figure 97

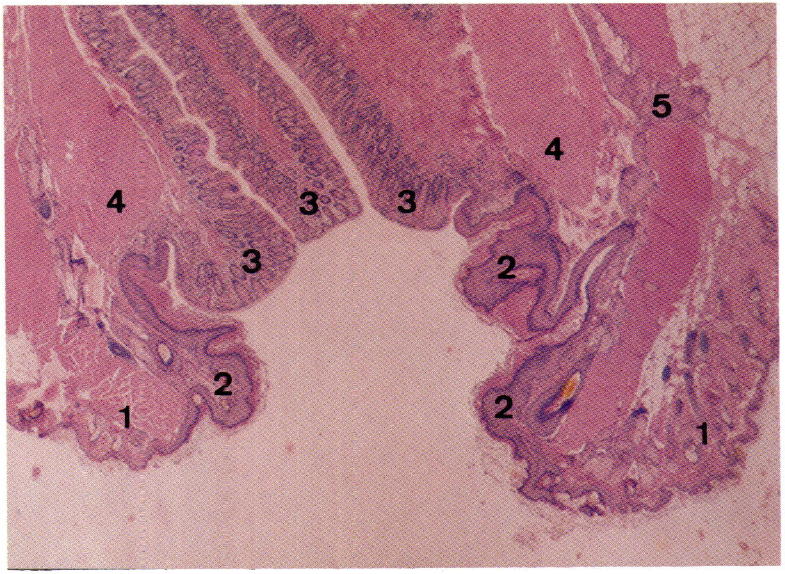

Figure 98

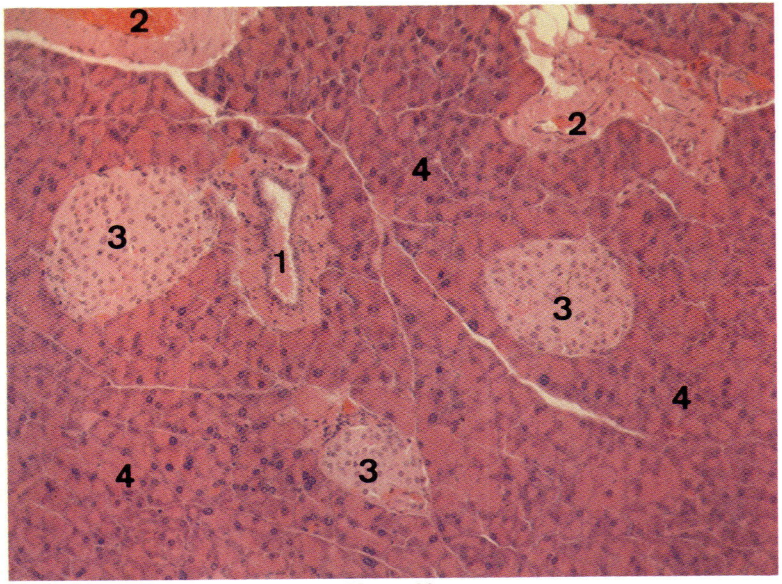

Figure 99

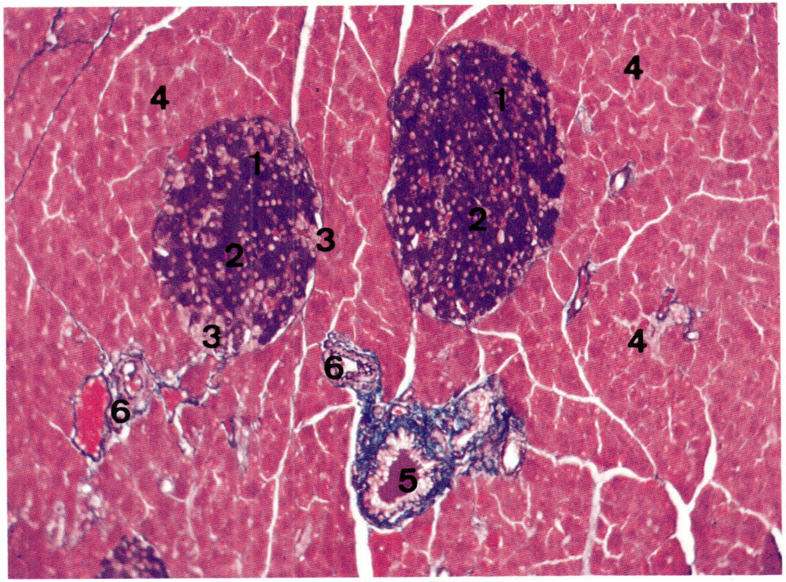

Figure 100

101. Liver — Heidenhain's Azan — 500×

Blue-stained reticulum forms the walls of sinusoids, spaces that contain erythrocytes. Fixed macrophages (Kupffer cells) are attached to the sinusoid wall. Hepatic cell nuclei are stained deep blue and cytoplasm is tan.

1. Hepatic cell
2. Sinusoid
3. Kupffer cell

102. Liver — H & E — 250×

The portal triad is surrounded by hepatocytes. In the center is the portal vein filled with blood. In the upper left corner adjacent to the vein is a branch of the hepatic artery. One branch of a bile duct lies at the upper and lower right areas.

1. Hepatocytes
2. Branch portal vein }
3. Bile duct } Portal triad
4. Hepatic artery }

103. Liver — H & E — 800×

The cytoplasm of these hepatic cells is filled with stored material rich in RNA. Each nucleus contains one or more large nucleoli in addition to chromatin granules. The large spaces between cords of hepatic cells are sinusoids; one Kupffer cell (fixed macrophage) is seen attached to the wall (upper right) and lies in the sinusoid. In the center between two hepatic cells is seen a small opening, a bile canaliculus.

1. Hepatocyte
2. Sinusoid
3. Nucleus—Kupffer cell
4. Opening—bile canaliculus

104. Gallbladder — H & E — 500×

The inner lining of the gallbladder is simple columnar epithelium. Beneath is a layer of loose connective tissue and blood vessels. A very thin layer of smooth muscle separates this area from serosa. Characteristic hepatic cells occupy the lower third of the section.

1. Simple columnar epithelium
2. Lamina propria
3. Blood vessel
4. Smooth muscle
5. Serosa
6. Hepatic cells

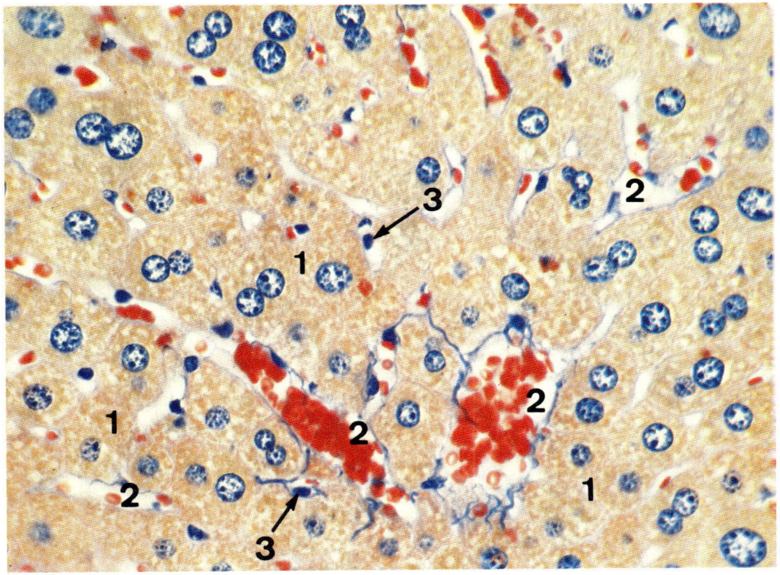

Figure 101

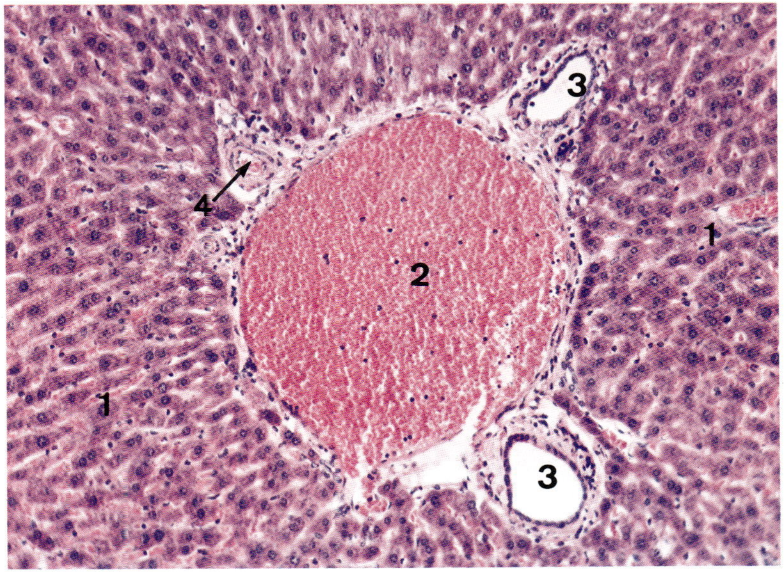

Figure 102

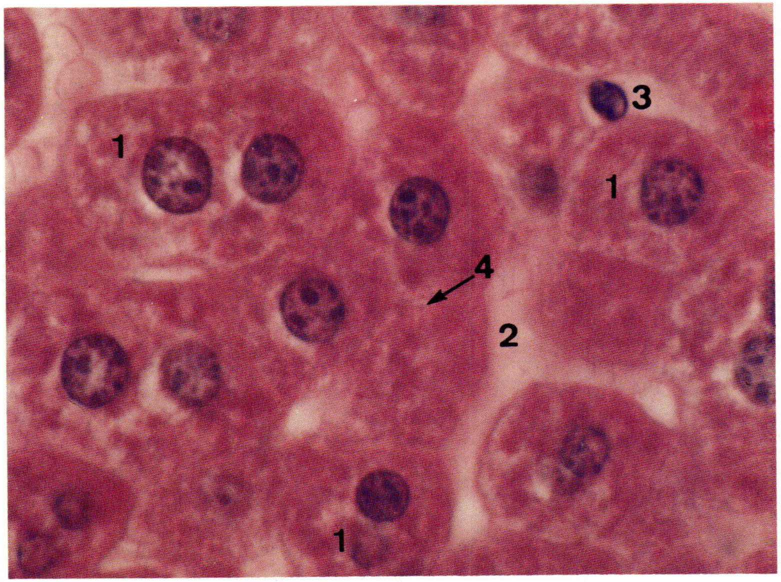

Figure 103

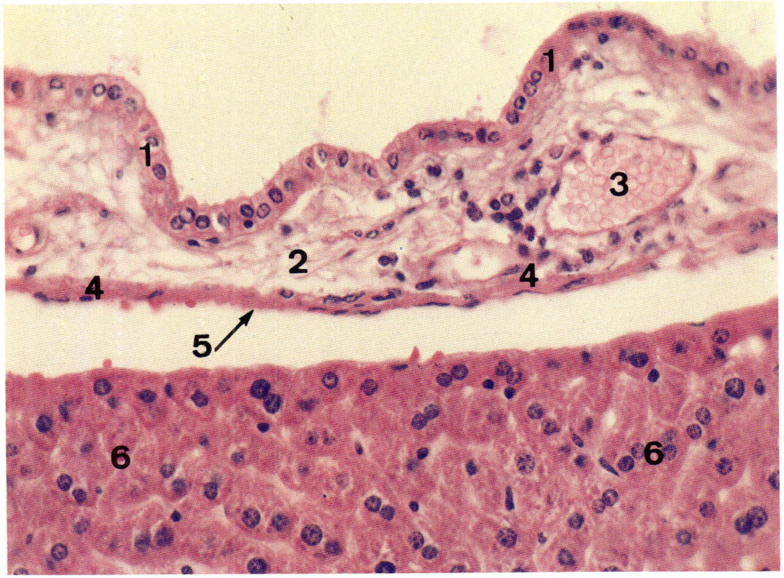

Figure 104

105. Kidney and adrenal H & E 50×

In this section the anatomical relationship between kidney and adrenal is clear. The cortex and medulla of the adrenal on the right are identified. In the kidney the cortex, with its proximal and distal convoluted tubules, may be compared with the longitudinal sections of collecting tubules in the medulla.

1. Adrenal medulla
2. Adrenal cortex
3. Kidney cortex
4. Kidney medulla

106. Kidney, male H & E 500×

Pronounced cuboidal epithelium of Bowman's capsule characterizes this tissue in the male. In the female the epithelial cells are much smaller.

1. Glomerulus
2. Cuboidal epithelium—Bowman's capsule
3. Proximal convoluted tubule
4. Distal convoluted tubule
5. Artery

107. Kidney tubules Heidenhain's Azan 500×

Proximal convoluted tubules with a blue brush border may be contrasted with distal convoluted tubules having no brush border. Mitochondria within the cytoplasm of tubular cells are stained brilliant red. A few tubules have dark blue casts within them, which may indicate minor damage to some glomeruli.

1. Proximal convoluted tubule (brush border) (blue)
2. Mitochondria (red) in tubular cells
3. Distal convoluted tubules
4. Cast in tubule (blue)

108. Kidney tubules Methacrylate Methylene Blue Basic fuchsin 800×

In this thin section the red mitochondria in the cytoplasm of tubular cells are clearly identified. Between the tubules in the lower right and central portions of the section are sections of blood vessels with erythrocytes. Endothelial cells line the blood vessels.

1. Nucleus of tubular cell
2. Mitochondria (red)
3. Endothelial cell
4. Erythrocytes
5. Nucleolus of tubular cell nucleus

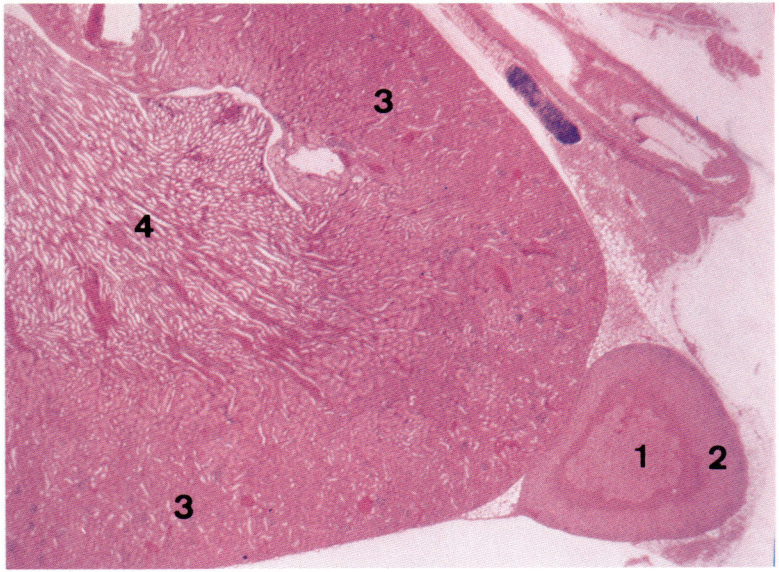

Figure 105

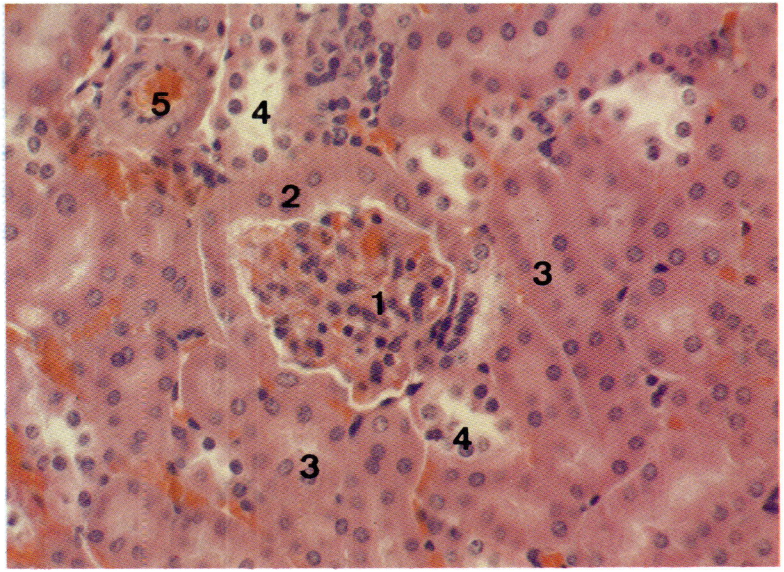

Figure 106

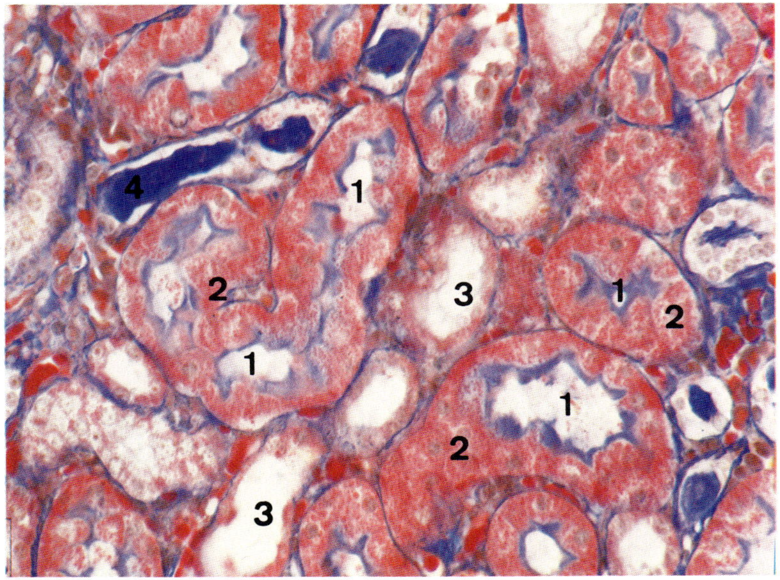

Figure 107

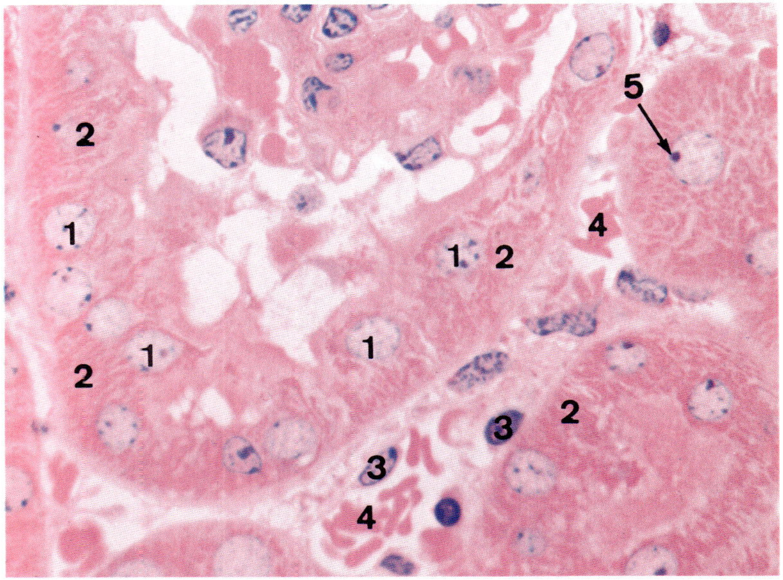

Figure 108

109. Bladder H & E 200×

Transitional epithelium (urothelium) lines the inside bladder wall; connective tissue fibers form the lamina propria. A thick wall of smooth muscle surrounds the bladder.

1. Transitional epithelium
2. Lamina propria
3. Blood vessel
4. Smooth muscle

110. Bladder Masson stain 250×

Green collagen fibers form most of the lamina propria underlying the transitional epithelium lining the bladder cavity. A lymphatic nodule is seen at upper left.

1. Transitional epithelium
2. Collagen—connective tissue
3. Lymphatic nodule

111. Urinary bladder H & E 250×

At higher magnification the characteristic morphology of transitional epithelium is apparent.

1. Transitional epithelium
2. Lamina propria
3. Smooth muscle

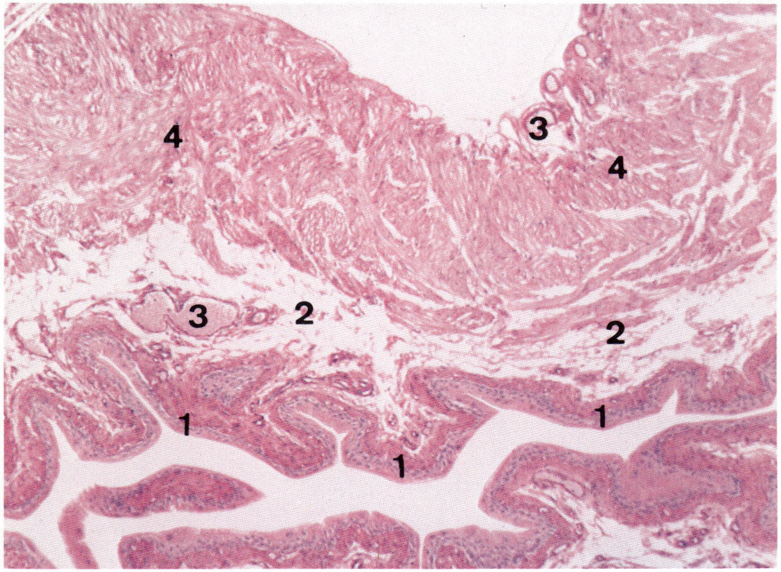

Figure 109

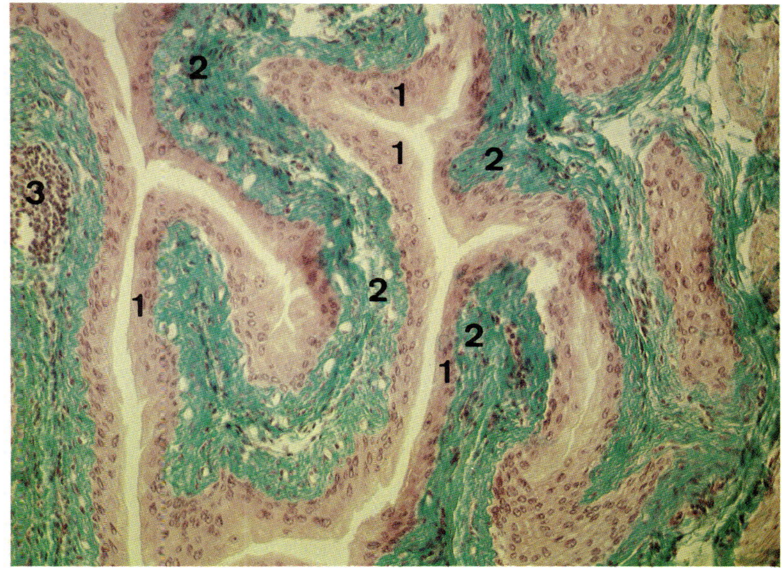

Figure 110

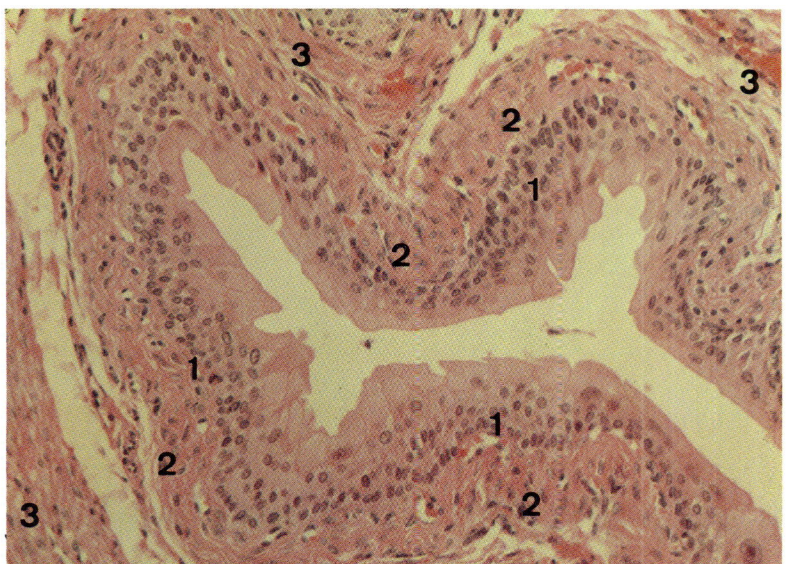

Figure 111

112. Connective tissue | Heidenhain's Azan | 500×

A clump of mast cells, each with blue cytoplasm and a red nucleus, is nestled among elastic and collagen fibers. Both types of fibers have taken the blue stain; elastic fibers are very thin and wavy.

1. Mast cell
2. Thin green elastic fiber
3. Collagen fibers (wavy green)

113. Connective tissue Mast cells | Hematoxylin, Aldehyde fuchsin, Fast Green | 500×

A group of mast cells with green nuclei and purple granules lies among thin, dark elastic fibers and green collagen fibers.

1. Mast cell
2. Thin, purple elastic fiber
3. Green collagen fibers
4. Macrophage
5. Fibroblast

114. Connective tissue | Hematoxylin, Aldehyde fuchsin, Fast Green | 500×

In this section of abdominal connective tissue fascia, thin, purple elastic fibers, green collagen fibers, elongated nuclei of fibroblasts, and oval nuclei of macrophages are clearly identified. Green-staining erythrocytes are in lower left.

1. Fibroblast
2. Elastic fiber (purple)
3. Wavy green collagen fibers
4. Erythrocytes

115. Mast cells | Methacrylate, Methylene Blue, Basic fuchsin | 250×

In the connective tissue of the dermis of the skin are several elongated fibroblasts and a row of tissue mast cells, below which are two empty fat cells. One medium artery is in upper left; the endothelial cells and smooth muscle cells are easily identified.

1. Mast cell (tissue basophil) blue granules
2. Fat cell
3. Artery
4. Endothelial cells
5. Fibroblast
6. Two smooth muscle cell nuclei

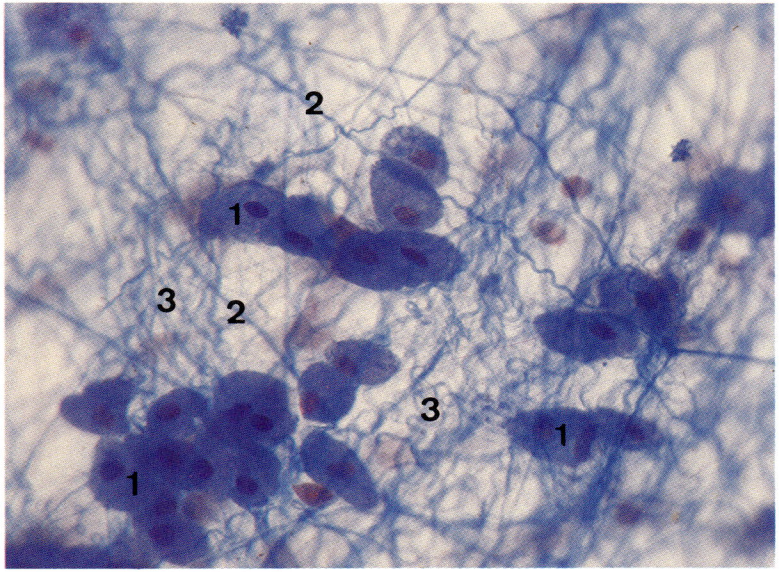

Figure 112

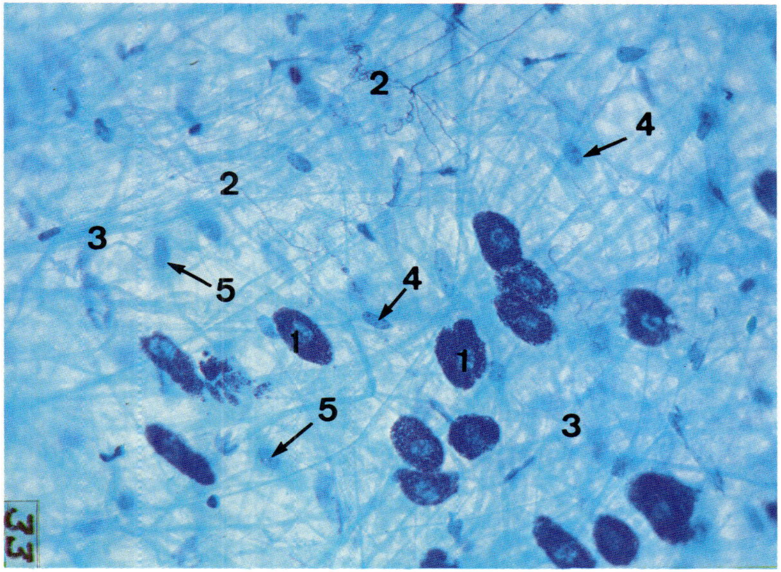

Figure 113

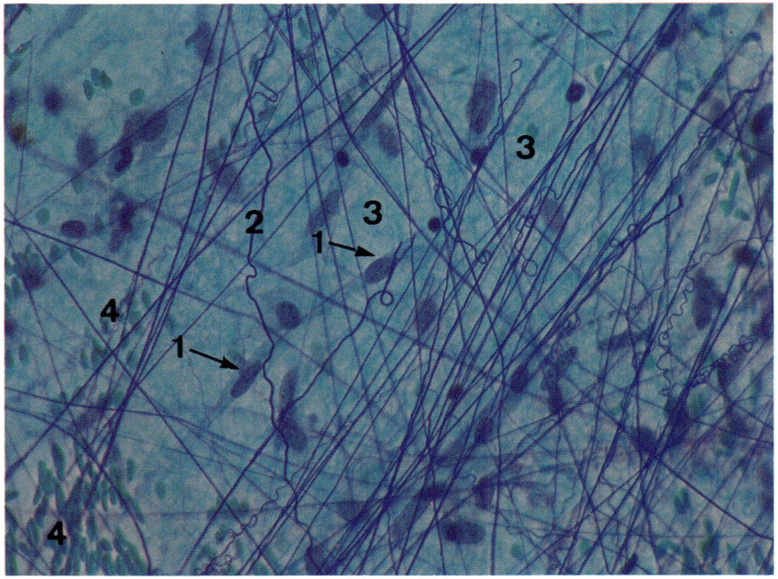

Figure 114

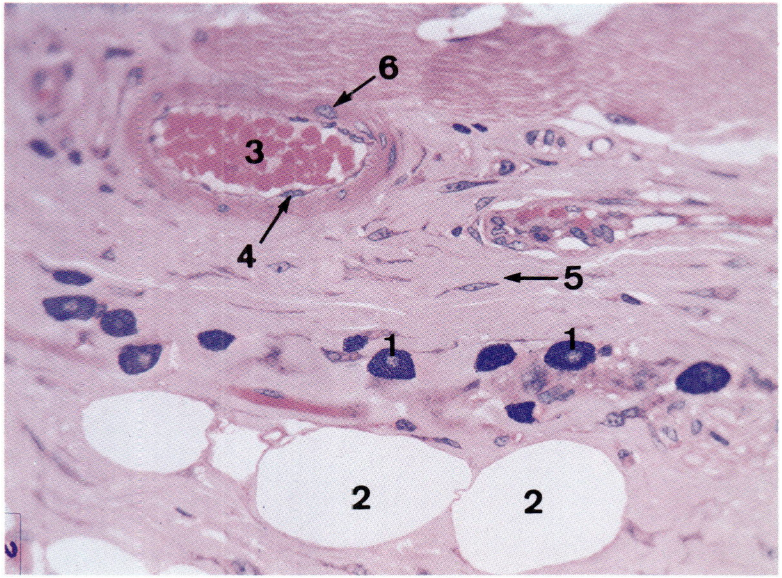

Figure 115

116. Hair bulb H & E 500×

This portion of the hair is in the dermis. Many fat cells surround the hair. One cell in the bulb region is undergoing division.

1. Connective tissue root sheath
2. Outer epithelial root sheath
3. Root of hair
4. Undifferentiated bulb cells
5. Blood vessel
6. Fat cell
7. Nucleus of fat cell
8. Cell in division

117. Hair follicle H & E 500×

This portion of the hair projects from the dermis through the epidermis to the skin surface. Keratin appears in the center of the hair. Sebaceous glands lie along the hair shaft.

1. Epidermis
2. Hair shaft
3. Hair follicle
4. Sebaceous gland

118. Vibrissae follicles H & E 50×

Vibrissae are very sensitive tactile hairs located on the lips and other areas of the face. They are larger than normal hair follicles and the root and bulb of the vibrissa are surrounded by nerve fibers and a rich blood supply.

1. Skin
2. Hair root
3. Nerve fibers
4. Pulvinus
5. Ring sinus with few erythrocytes
6. Hair follicle
7. Striated muscle

119. Vibrissa follicle Masson stain 50×

Similar structures identified in the H & E section are seen with the Masson stain. The hair shaft is bright red. Green collagen fibers occur in the connective tissue sheath and hyaline membrane. Blood is present in the ring sinus.

1. Hair bulb
2. Papilla
3. Nerve fibers
4. Pulvinus
5. Connective tissue sheath
6. Ring sinus
7. Hyaline membrane
8. Sebaceous gland
9. Epidermis
10. Hair shaft
11. Fat cells
12. Striated muscle

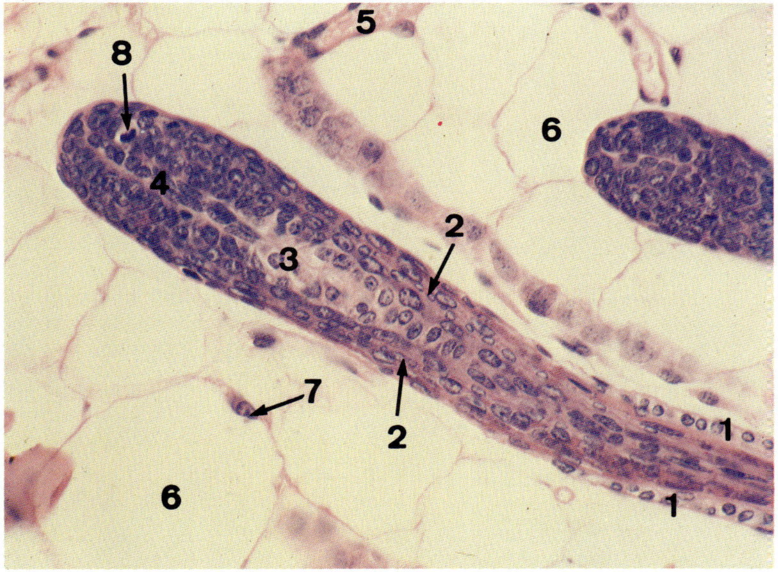

Figure 116

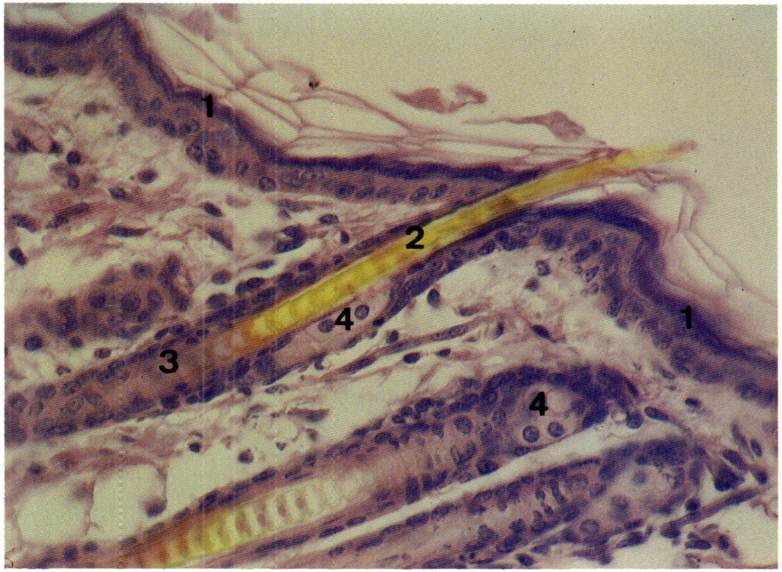

Figure 117

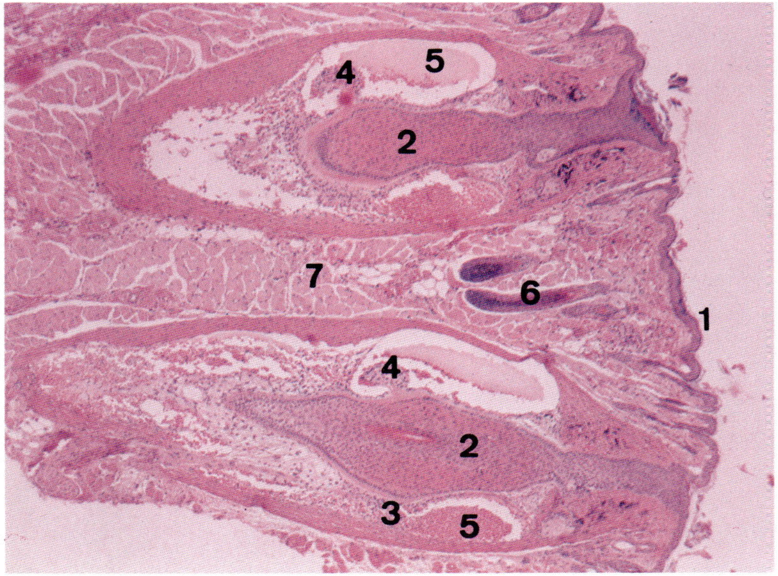

Figure 118

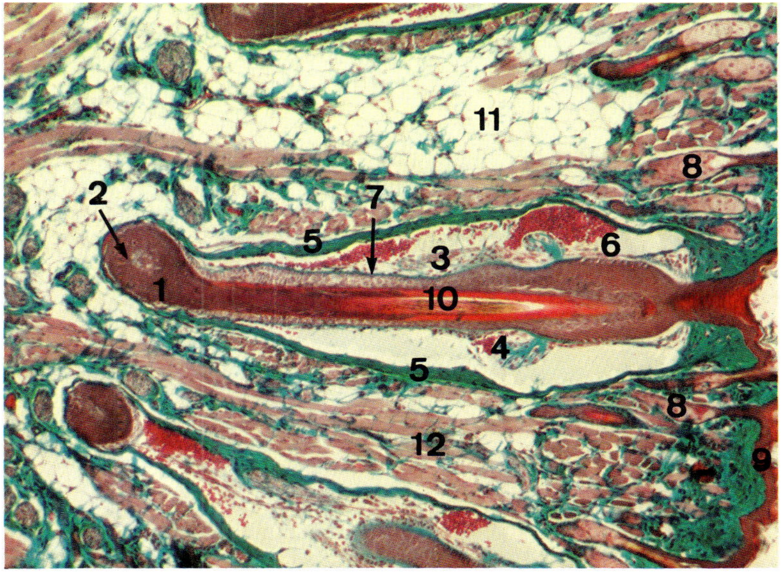

Figure 119

120. Mucous gland and vessels — Aldehyde–fuchsin, Masson stain — 100×

The mucous gland shown centrally stains purple. One large and two medium-sized arteries with prominent internal elastic membranes are seen at right and at center left. Green collagen fibers are present in the tunica media and adventia. Brown striated muscle fibers are at lower right. Two nerve trunks in cross section are located between the mucous gland and large artery.

1. Purple mucous secretion
2. Internal elastic membrane (artery)
3. Erythrocytes in artery
4. Nerve trunk
5. Striated muscle

121. Thick skin, Sole of foot — H & E — 500×

The outer layer of flattened dead cells represents the stratum corneum. Keratohyaline granules are present in the underlying stratum granulosum. The next layer of four to five cells forms the stratum germinativum. Connective tissue of collagen and elastic fibers plus blood and lymphatic vessels and nerves comprise the dermis. Two dermal papillae containing blood vessels and nerves extend up into the epidermis at center and left.

1. Dermis connective tissue
2. Dermal papillus
3. Stratum germinativum
4. Stratum granulosum
5. Stratum corneum
6. Desquamating keratin

122. Fat: Brown and white — H & E — 250×

Brown fat is easily differentiated from white by the central position of the nucleus in the cell. White fat cells appear as spaces whose nuclei are located at the periphery of the cell.

1. Brown fat cells
2. White fat cells
3. Erythrocytes in vein
4. Small arteriole

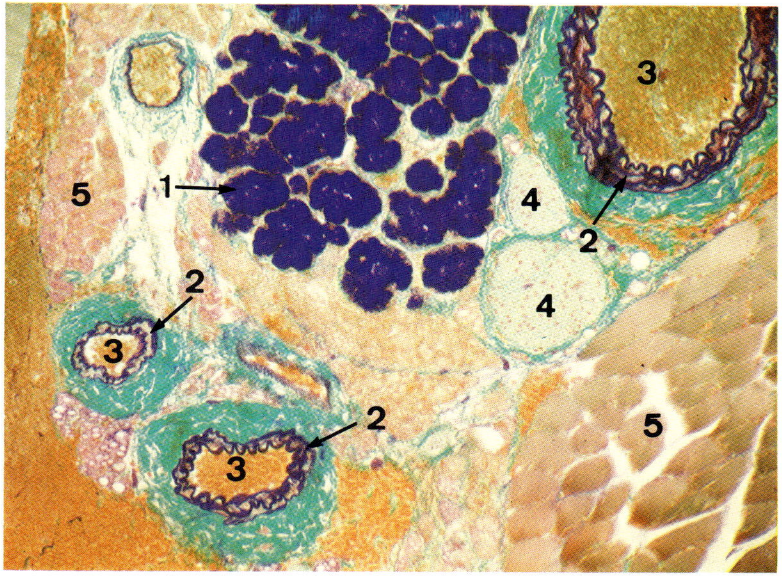

Figure 120

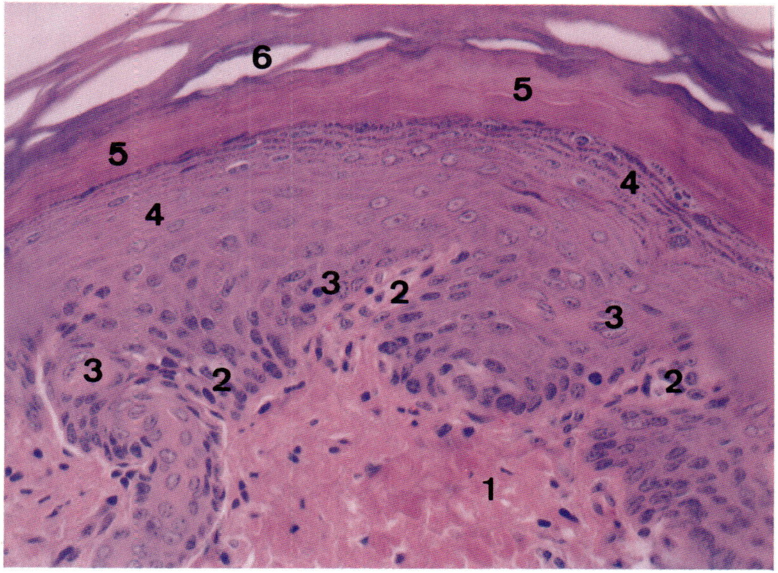

Figure 121

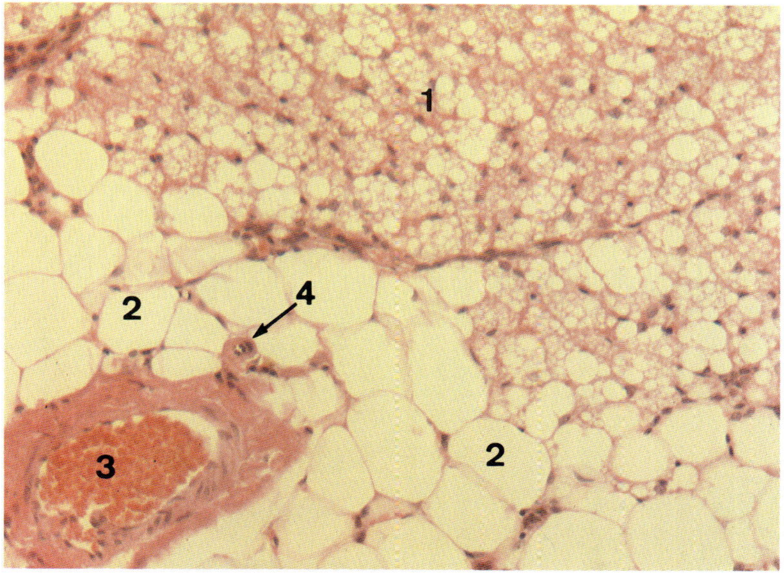

Figure 122

123. Junction: Nasal passage, trachea, esophagus — H & E — 50×

At low magnification the nasal passage curves into the trachea at top center. The tracheal lumen curves to the left. Hyaline cartilage in the tracheal wall supports it and keeps it from collapsing. The epiglottis is seen covering the esophageal opening at center right. Bone and striated muscle are seen in the lower third. Bony vertebrae are located at the top.

1. Nasal passage
2. Mouth
3. Epiglottis
4. Trachea
5. Esophagus
6. Soft palate
7. Vertebra
8. Tracheal cartilage
9. Pharynx
10. Striated muscle

124. Nasal mucosa — Masson stain — 100×

Ciliated columnar epithelium with goblet cells lines the sinuses of the nose as it passes around an island of hyaline cartilage and forms the characteristic specialized olfactory epithelium with serous and mucous glands.

1. Hyaline cartilage
2. Ciliated columnar epithelium
3. Serous and mucous glands with goblet cells
4. Sinuses

125. Nasal mucosa — H & E — 500×

The nasal mucosa is lined by pseudostratified ciliated columnar epithelium with goblet cells. Tall sustentacular cells with oval nuclei lie just beneath the epithelium. Olfactory cells (bipolar ganglion cells) with dark spherical nuclei lie in the deepest part of the mucosa and send axons into the underlying connective tissue to form olfactory nerve trunks. Bone forms a boundary through which a blood vessel passes.

1. Pseudostratified ciliated columnar epithelium
2. Sustentacular cells
3. Mucous glands
4. Bone
5. Erythrocytes in blood vessels
6. Olfactory cells

126. Trachea — Masson stain — 250×

Ciliated columnar epithelium with goblet cells lines the trachea. Beneath the epithelium are serous and mucous glands with blood vessels and nerves lying immediately adjacent to hyaline cartilage.

1. Ciliated columnar epithelium
2. Serous gland
3. Mucous glands
4. Hyaline cartilage
5. Goblet cell

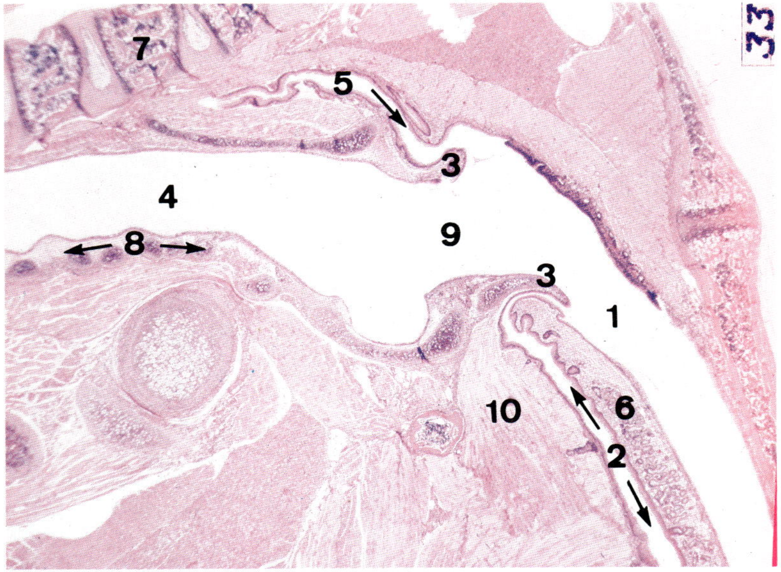

Figure 123

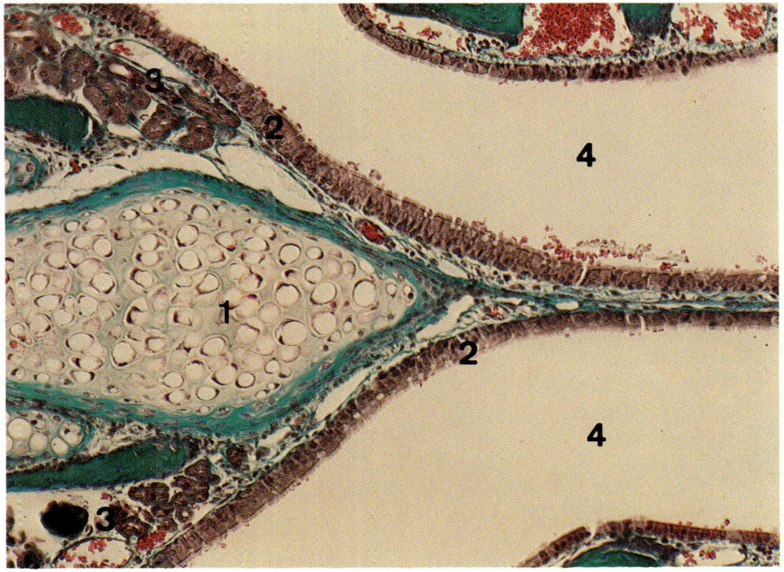

Figure 124

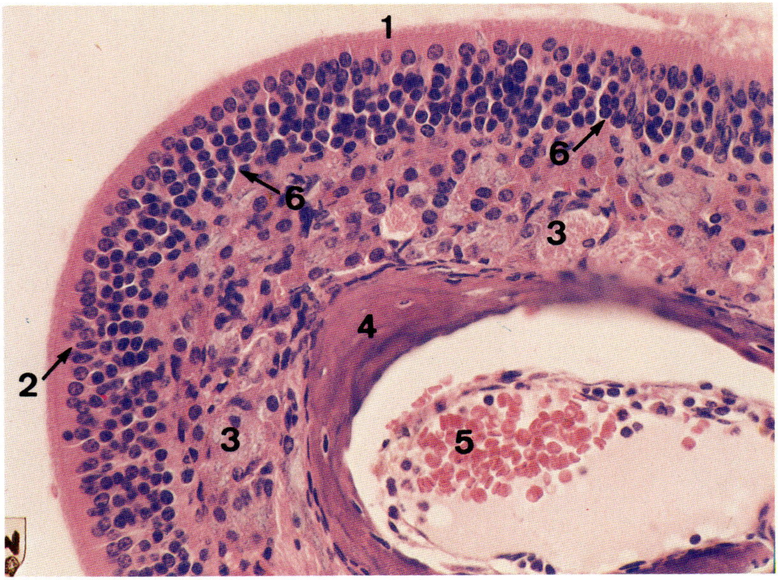

Figure 125

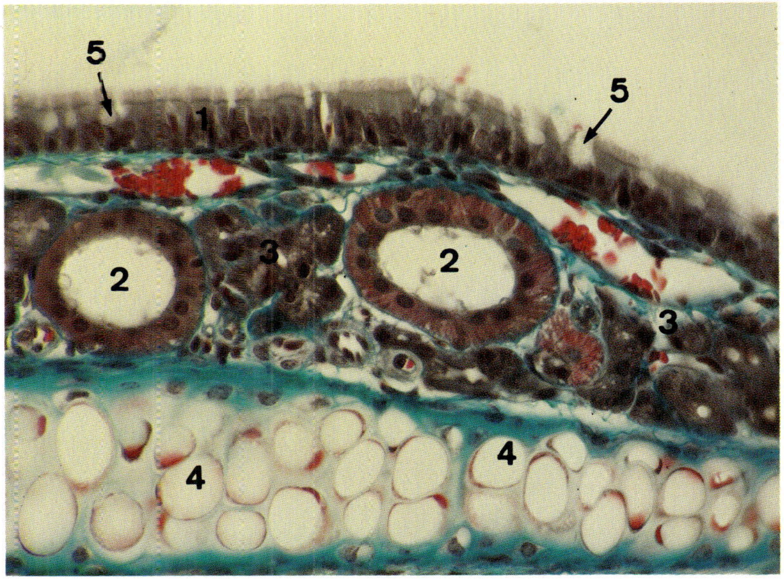

Figure 126

127. Lung H & E 100×

A terminal bronchiole ending in an alveolar duct in upper center. Small cavities opening on this duct are alveolar sacs. These are surrounded by alveoli.

1. Terminal bronchiole
2. Alveolar duct
3. Alveolar sac
4. Alveolus

128. Lung Methacrylate Toluidine Blue 250×

The lumen of a respiratory bronchiole is lined by low columnar epithelium. The spongy appearance arises from alveolar ducts and sacs in the respiratory tissue.

1. Alveolus
2. Alveolar sac
3. Bronchiole
4. Blood vessel
5. Alveolar duct

129. Lung Snook's Silver Stain Nuclear Fast Red 250×

Black reticular fibers occur within connective tissue of respiratory tract.

1. Reticular fibers (black)
2. Alveolar duct
3. Alveolar sac

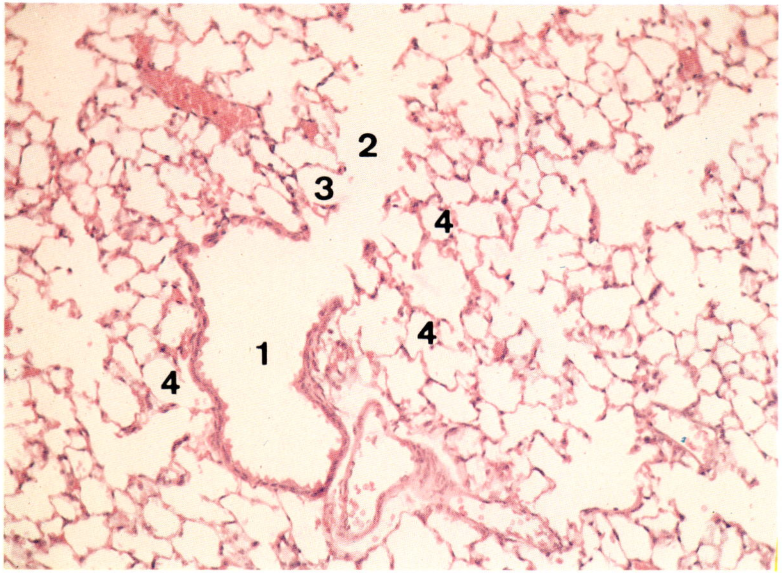

Figure 127

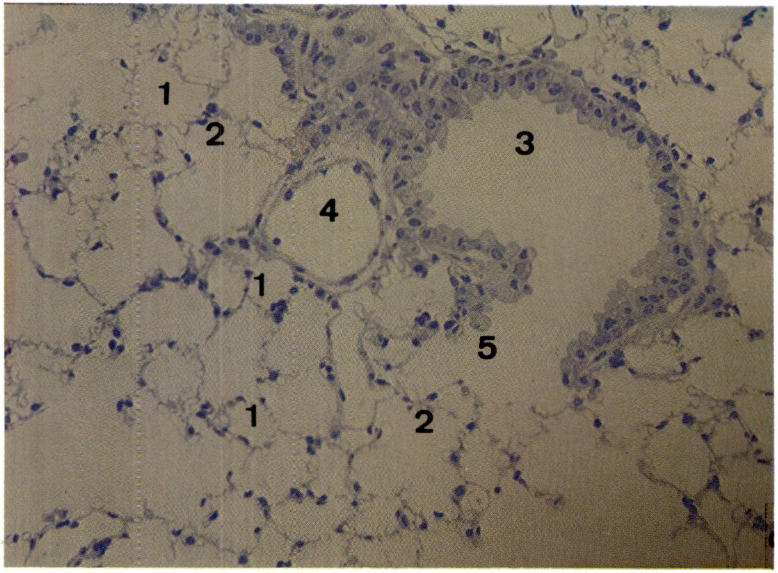

Figure 128

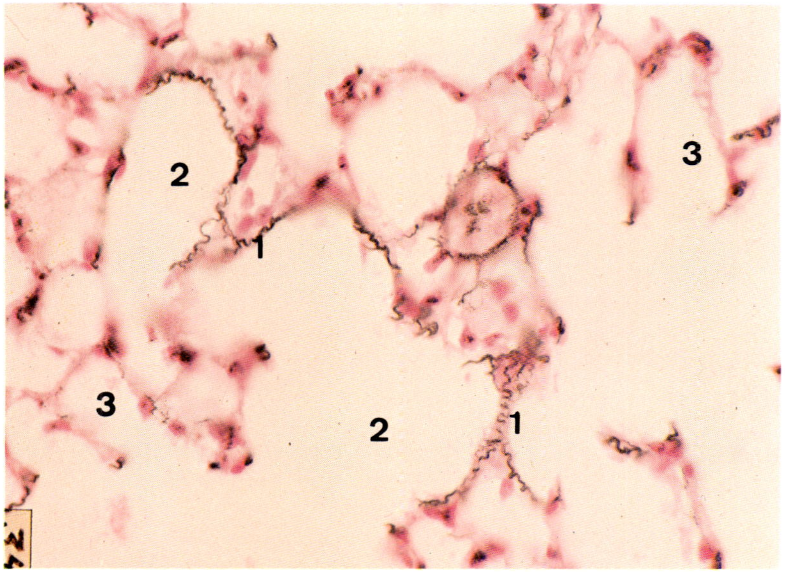

Figure 129

130. Lung

Aldehyde–fuchsin
Hematoxylin
Fast Green

500×

Abundant purple elastic fibers are seen throughout the lung tissue.

1. Alveolus
2. Alveolar sac
3. Bronchiole
4. Elastic fiber (purple)
5. Blood vessel

131. Bronchiole

Methacrylate
Methylene Blue
Basic fuchsin

500×

In this 1.5-μm methacrylate section of bronchiole lining, ciliated cells may be distinguished from nonciliated Clara cells.

1. Ciliated cell
2. Clara cell
3. Connective tissue
4. Alveolus

132. Lung

Methacrylate
Toluidine Blue

500×

Type I and II alveolar cells can be distinguished clearly in this 1.5-μm section.

1. Type II (great alveolar) cell
2. Type I cell
3. Endothelial cell
4. Capillary
5. Alveolus
6. Septum

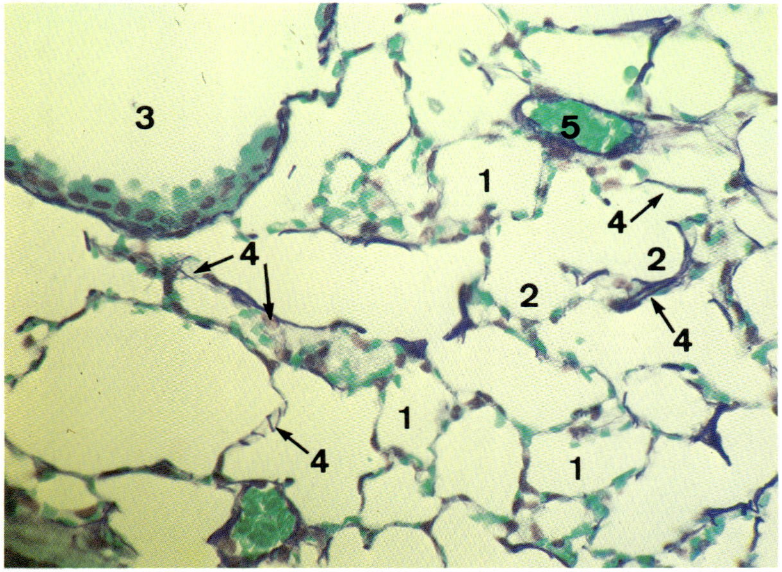

Figure 130

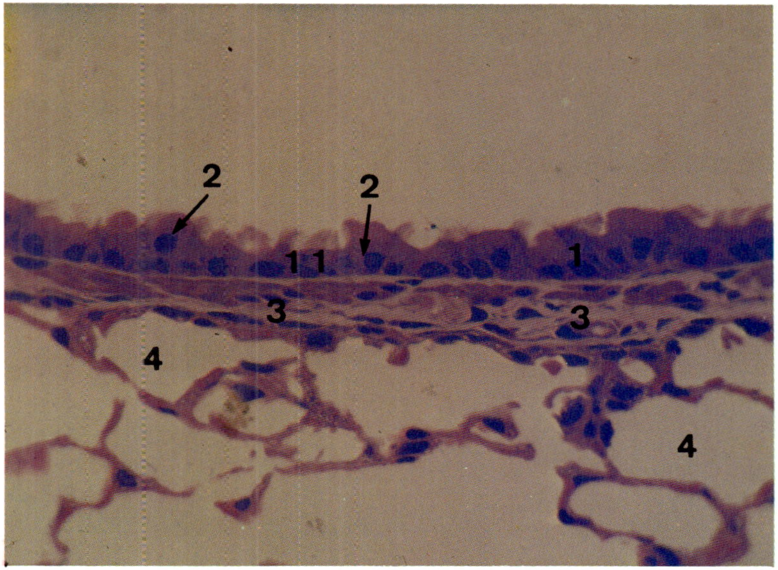

Figure 131

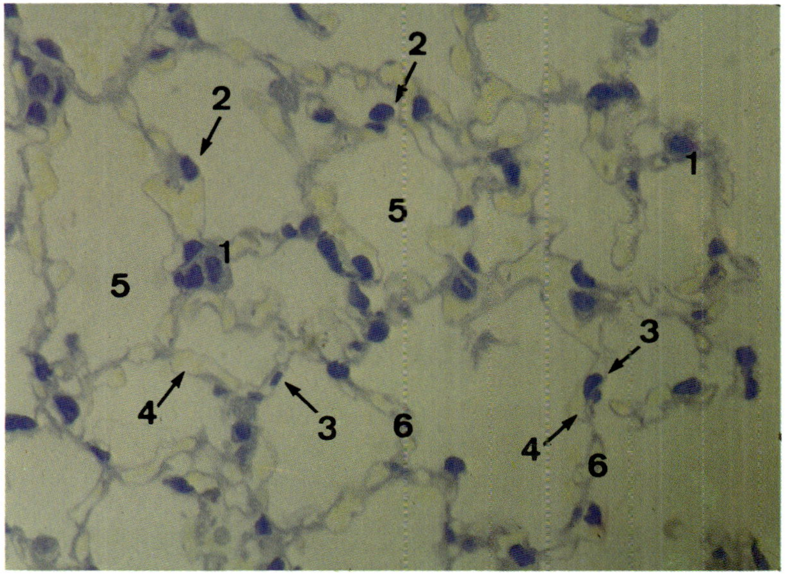

Figure 132

133. Thymus H & E 50×

The very dark cortex consists of masses of small lymphocytes. Bands of pink trabeculae originate in the capsule and penetrate the cortex. The lighter medulla is located centrally. Large blood vessels as well as masses of fat cells are located in the margins.

1. Cortex
2. Medulla
3. Fat cells
4. Artery
5. Arterial wall
6. Striated muscle
7. Trabeculae

134. Thymus H & E 500×

Differentiation of dark-staining cortex and lighter medulla can be made easily, with several groups of large, pale epithelial cells forming Hassall's corpuscle in the medulla.

1. Cortex
2. Medulla
3. Hassall's corpuscle

135. Spleen H & E 50×

Dense connective tissue forms a tough capsule surrounding the spleen. Extending inward from the capsule are bands of trabeculae, composed of reticular fibers and smooth muscle, which form a framework for white and red pulp. Nodules of lymphocytes form white pulp, which contains a central arteriole. The red pulp is composed of erythrocytes, granular leukocytes, lymphocytes, and megakaryocytes.

1. Capsule
2. Germinal center
3. Trabeculum
4. Red pulp
5. White pulp

136. Spleen H & E 100×

At a higher magnification, structures in the red and white pulp are more easily seen.

1. Lymphoid nodule
2. Central arteriole
3. Trabeculum
4. Megakaryocyte
5. Red pulp

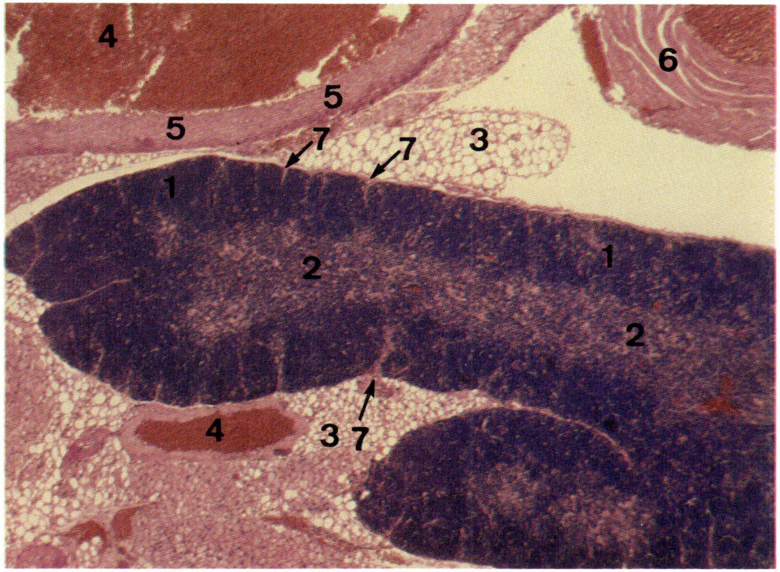

Figure 133

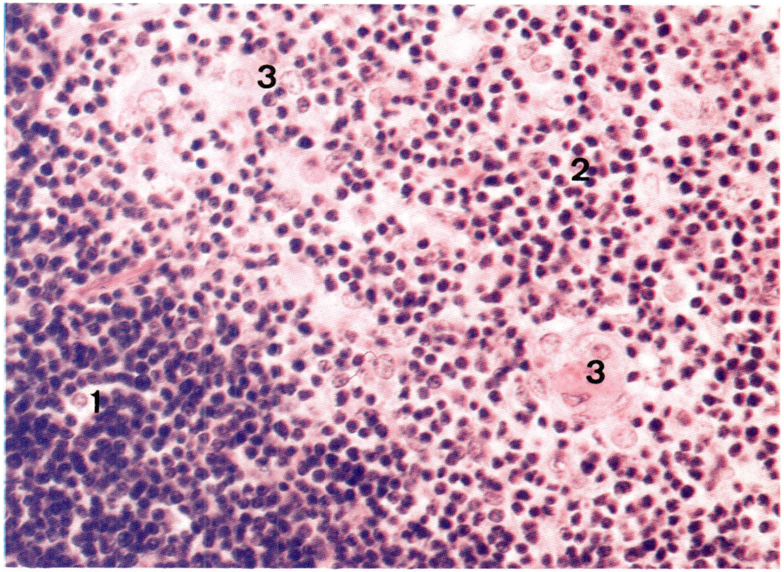

Figure 134

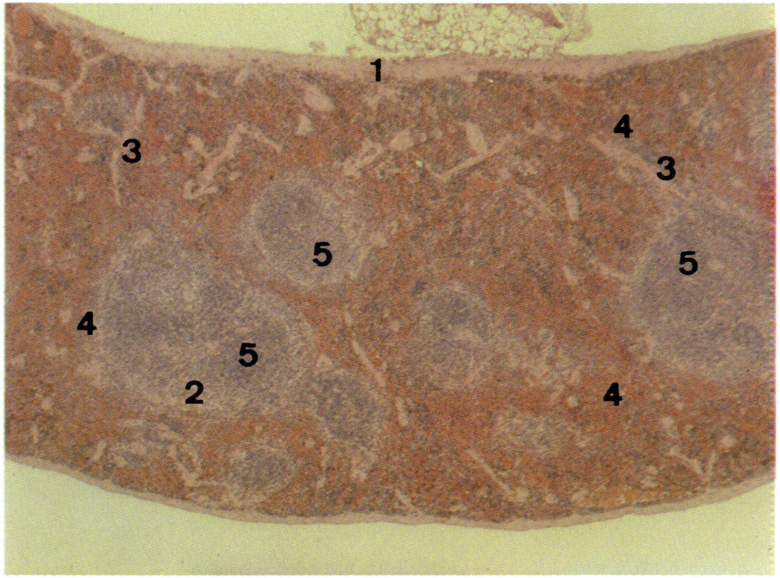

Figure 135

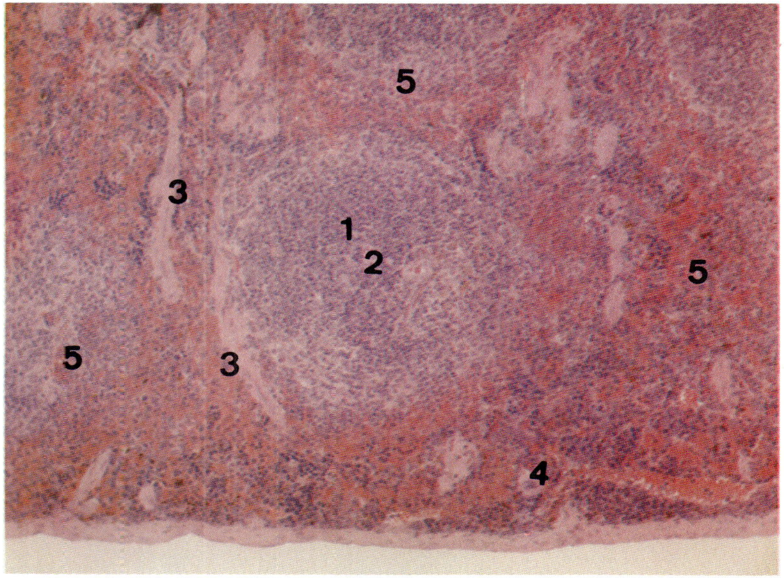

Figure 136

137. Spleen H & E 500×

In this magnified section five megakaryocytes in the red pulp are present in the left and center areas. Aggregations of erythrocytes in various stages of maturity surround them. At lower right lymphocytes form a portion of a nodule in white pulp.

1. Capsule
2. Trabeculum
3. Megakaryocyte
4. Red pulp

138. Lymph node H & E 50×

The lymph node, with its nodules, sinuses, and blood vessels, is surrounded by portions of the salivary glands.

1. Lymphoid follicle
2. Hilus
3. Sublingual gland
4. Submandibular gland
5. Intercommunicating sinus
6. Blood vessel
7. Paracortical area
8. Peripheral and marginal sinuses

139. Lymph node H & E 250×

A germinal center is seen in the upper left area of the node. The dark paracortical area on the right contains lymphocytes. Fat cells lie between the lymph node and blood vessel on the left.

1. Germinal center in lymphoid follicle
2. Lymphocytes in node
3. Artery
4. Adipose tissue

140. Spleen Snook's Silver Stain, Nuclear Fast Red 100×

A well-defined capsule surrounds the spleen. Trabeculae of connective tissue, blood vessels, and black reticular fibers extend from the capsule into the spleen proper. One large follicle of white pulp is in right center, surrounded by red pulp. Two megakaryocytes are in the red pulp at upper left.

1. Capsule
2. Trabeculum—black reticular fibers
3. Malpighian corpuscle—white pulp
4. Red pulp
5. Megakaryocytes

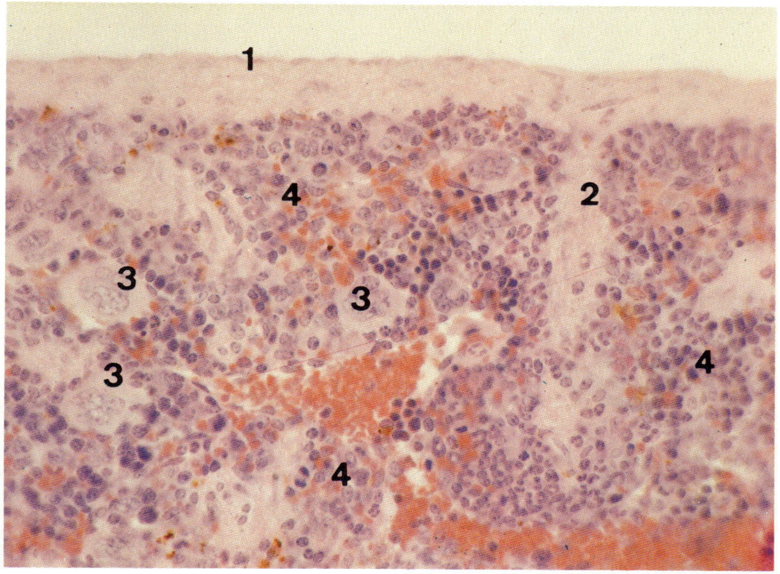

Figure 137

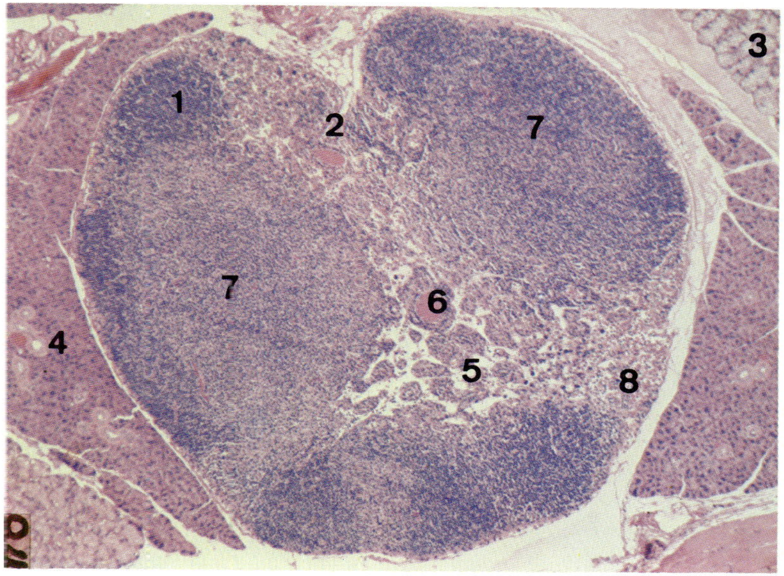

Figure 138

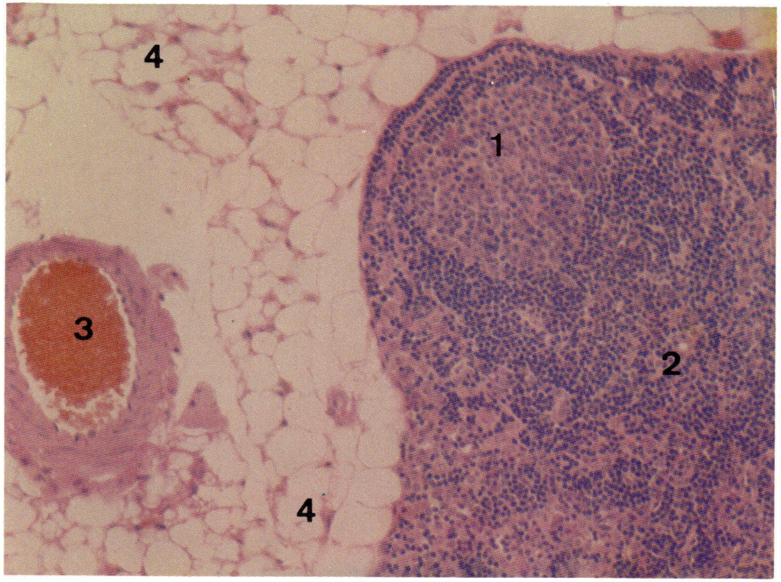

Figure 139

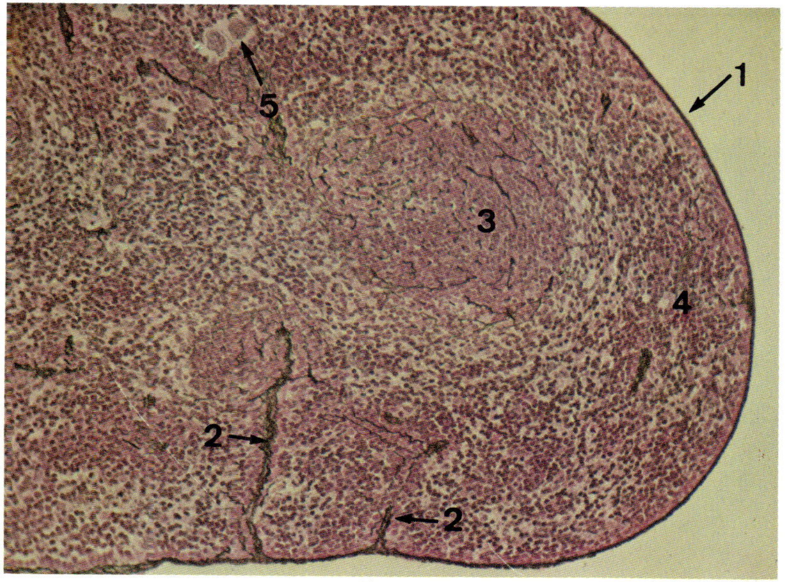

Figure 140

141. Cartilage H & E 500×

Rows of oval cells arranged in lines on the right are those of fibrocartilage. Large oval cells not so arranged but occupying the lower center are cells of hyaline cartilage. Elongated cells adjacent to the edges of bone are osteoblasts. A few erythrocytes are present in center.

1. Osteoblasts
2. Fibrocartilage
3. Hyaline cartilage
4. Bone

142. Bone H & E 500×

Two osteocytes are present in the trabeculae in the center of the field. On the right is an indentation, Howship's lacuna, with an osteoclast immediately adjacent. Bone marrow cells fill the area on the right.

1. Osteocyte
2. Osteoblasts
3. Osteoclast
4. Bone marrow cells
5. Howship's lacuna
6. Bone

143. Fibrocartilage H & E 150×

Fibrocartilage in the vicinity of a joint is attached to bone and tendon.

1. Fibrocartilage
2. Bone
3. Tendon attached to hyaline cartilage
4. Striated muscle

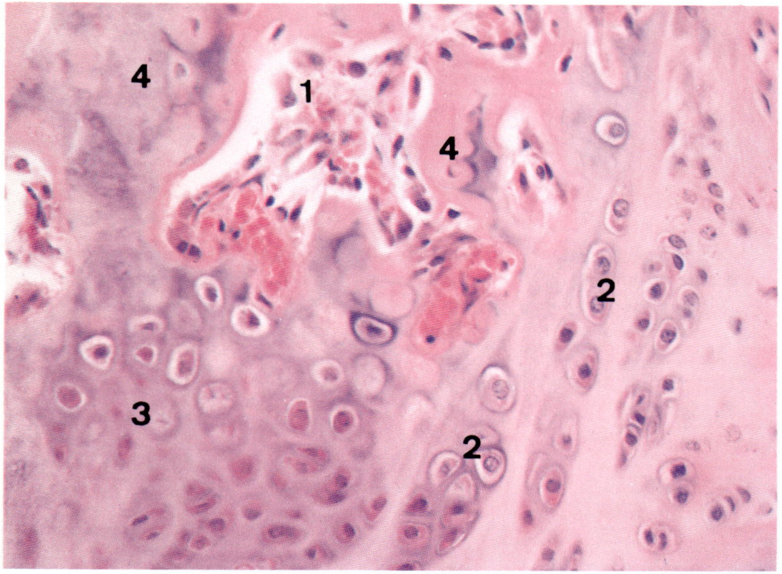

Figure 141

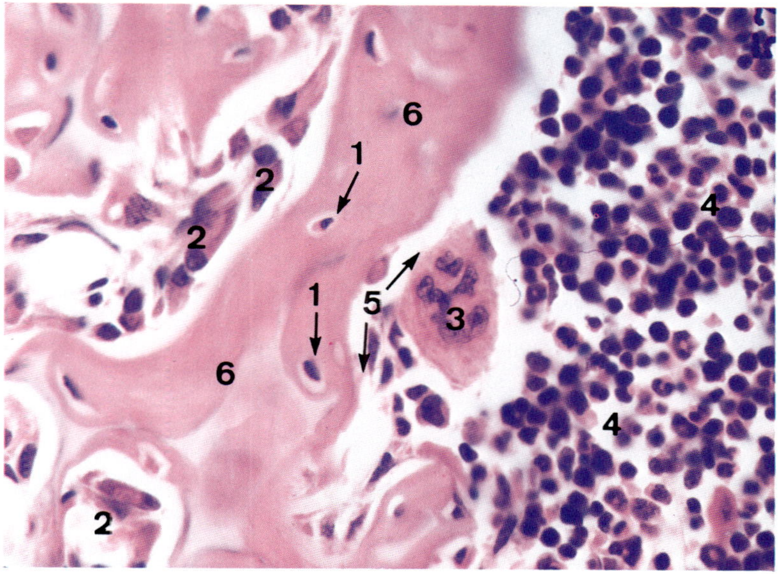

Figure 142

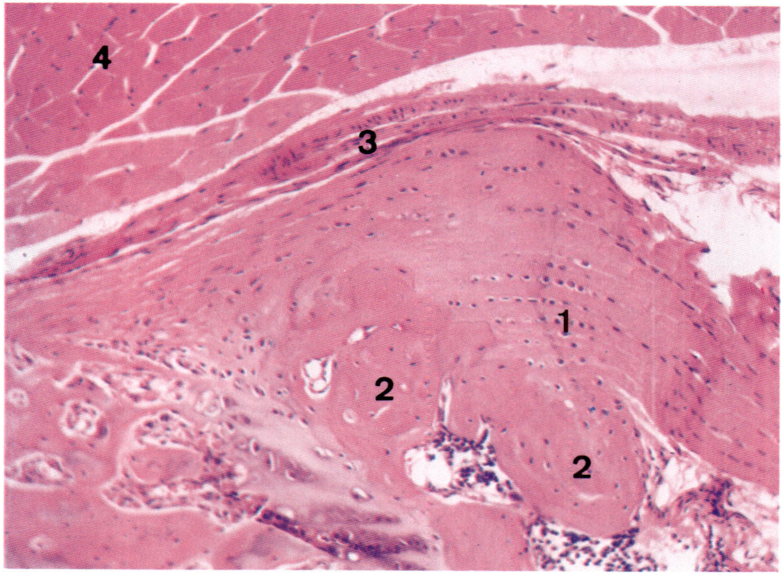

Figure 143

144. Femur: Epiphyseal plate H & E 200×

This section from a 1-month-old mouse shows the developing zones of bone formation from cartilage at right to bone marrow cells at left of the field.

1. Zone of proliferation
2. Zone of maturation
3. Zone of hypertrophy
4. Trabeculae
5. Bone marrow cells

Terminology from Bloom and Fawcett (1968).

145. Knee joint H & E 50×

The articular surfaces of bone forming the joint are of hyaline cartilage, below which is bone. Marrow cells are seen as dark masses between bone spicules. Above, below, and in between the articular surfaces are triangular structures of bone, fibrocartilage, and blood vessels—the menisci.

1. Articular cartilage (hyaline)
2. Meniscus
3. Bone marrow
4. Bone
5. Trabeculae (bone light; cartilage dark)
6. Striated muscle

146. Tail: Cross section H & E 150×

Stratified squamous epithelium forming the epidermal covering completely surrounds the tail. The four large pink oval structures located in each of the four quadrants are tendons. The central homogeneous red structure surrounding spaces is bone. Very few marrow cells are found in the spaces between the bone.

1. Stratified squamous epithelium
2. Tendon
3. Marrow cavity with very few cells
4. Hair follicle

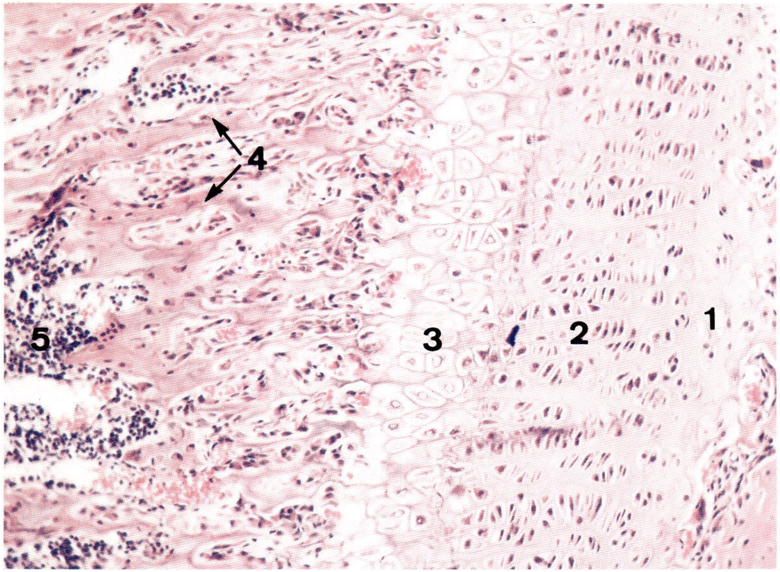

Figure 144

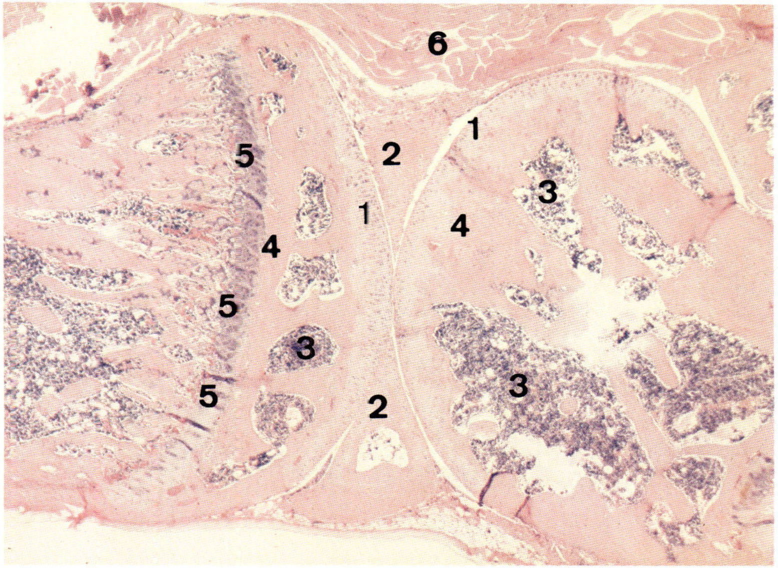

Figure 145

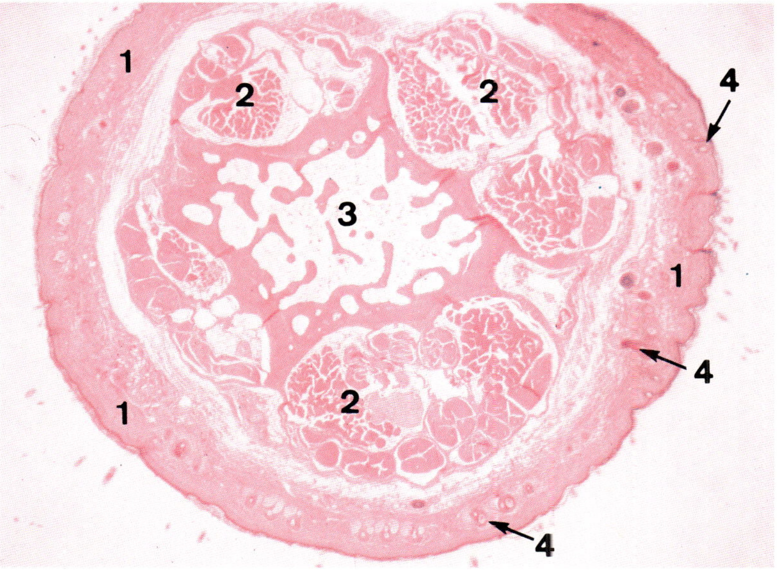

Figure 146

147. Incisor tooth H & E 50×

The incisor tooth projects from the lower jaw, showing the structures above and below the gum line.

1. Enamel
2. Dentine
3. Gingiva (gum)
4. Alveolar bone
5. Pulp cavity

148. Tooth H & E 250×

In this section the enamel has been removed and the outer layer is dentine. Extending perpendicularly through this layer are dentinal tubules, minute canals that originate from the pulp cavity and penetrate all parts of the dentine. A row of odontoblasts is prominent between the dentine and the pulp cavity. Macrophages, lymphocytes, blood vessels, and connective tissue fibers comprise the pulp. An acellular cementum covers the dentine below the gum line.

1. Dentine with dentinal tubules (striations)
2. Odontoblasts
3. Dental pulp
4. Acellular cementum

149. Tooth H & E 250×

The outermost layer of tissue is the stellate reticular layer of enamel pulp. Subjacent are the ameloblast cells, which aid in enamel formation. Below is a wide band of enamel that covers the dentine layer. A row of odontoblasts is seen between the dentine and dental pulp. The latter is a connective tissue layer with macrophages and lymphocytes scattered throughout.

1. Enamel
2. Dentine
3. Odontoblasts
4. Dental pulp
5. Ameloblast layer
6. Enamel pulp

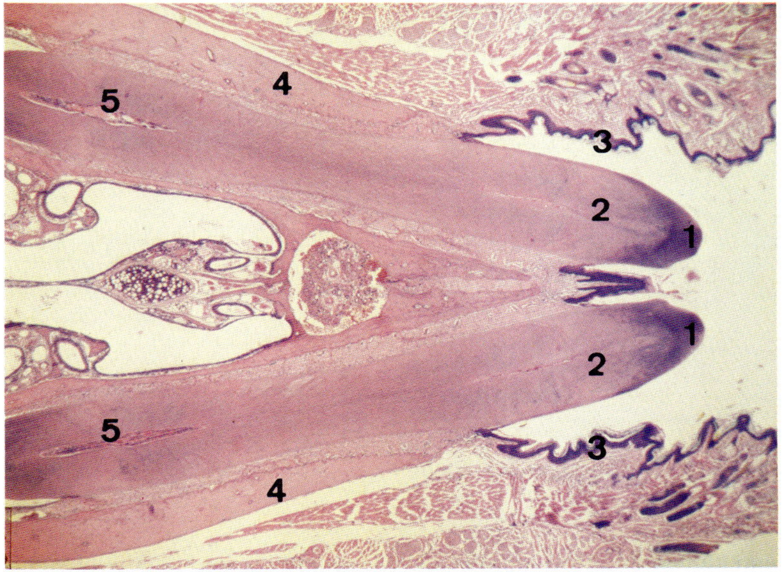

Figure 147

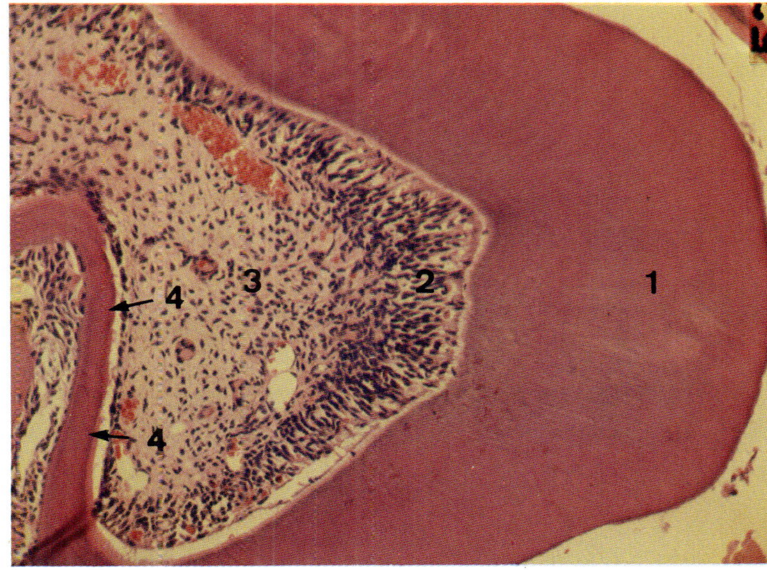

Figure 148

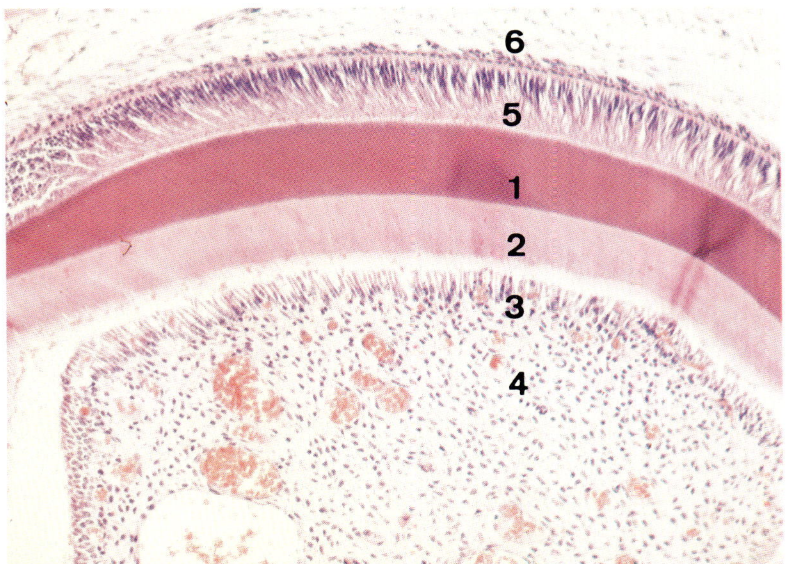

Figure 149

150. Striated muscle — Masson stain — 500×

The vertical striations and location of muscle nuclei at the side of the bundle of fibers characterize this type of muscle. Because this section is 5 μm thick, muscle fibers overlap each other and their nuclei do not always lie at the edge of the muscle bundle.

1. Artery
2. Collagen fibers (green)
3. Muscle fiber nucleus
4. Muscle fibers with striations

151. Skeletal muscle — Heidenhain's Azan — 500×

Characteristic vertical cross lines along the muscle fibers readily identify this type of muscle. Oval nuclei between muscle bundles are muscle nuclei. Other round nuclei are those of wandering leukocytes. Abundant collagen, elastic, and reticular fibers form a connective tissue, conveying blood, lymph vessels, and nerves and forming a boundary between muscle bundles.

1. Vertical striations along muscle bundle
2. Muscle fiber nucleus
3. Leukocytes in connective tissue
4. Blood vessel
5. Blue collagen fibers

152. Smooth muscle — H & E — 500×

Long spindle-shaped fibers are arranged in parallel. The oval, elongated nuclei lie in the middle of each fiber.

1. Muscle fiber nucleus
2. Muscle fiber
3. Blood vessel

153. Cardiac muscle — Heidenhain's Azan — 500×

The presence of nuclei in the center of muscle bundles and cross striations characterize this type of muscle. Another identifying feature is the division between sarcolemma junctions forming intercalated disks, stained blue with this stain. Blood and lymph vessels and nerves accompany blue collagen fibers between muscle bundles.

1. Muscle fiber nucleus
2. Intercalated disk
3. Blood vessel

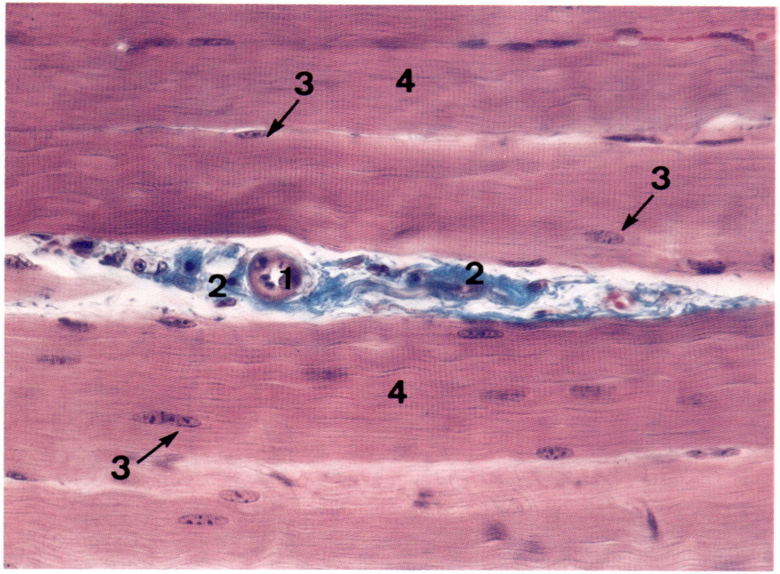

Figure 150

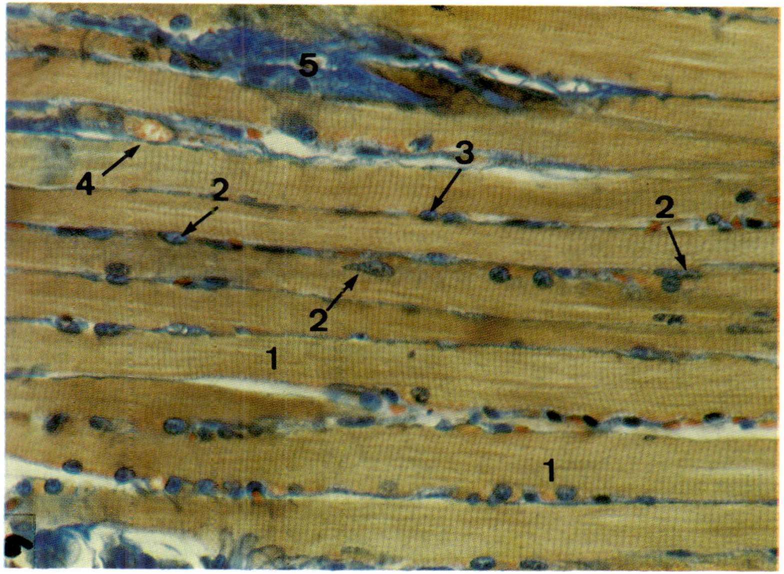

Figure 151

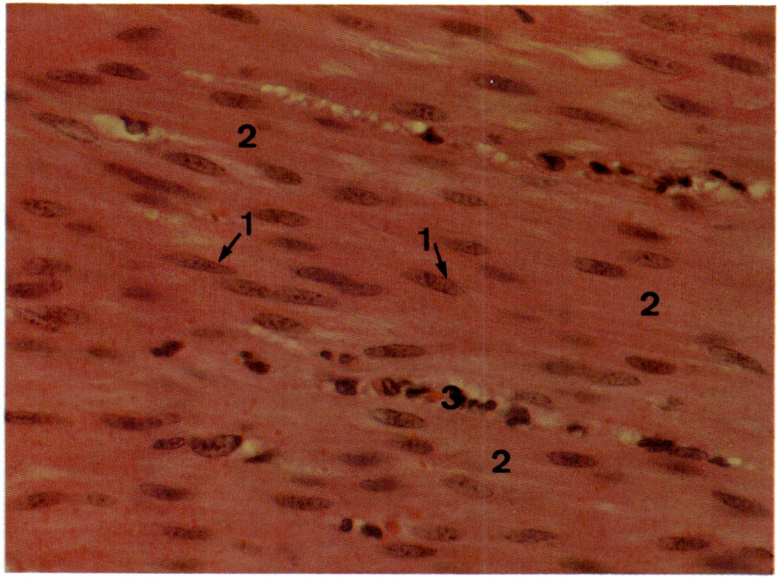

Figure 152

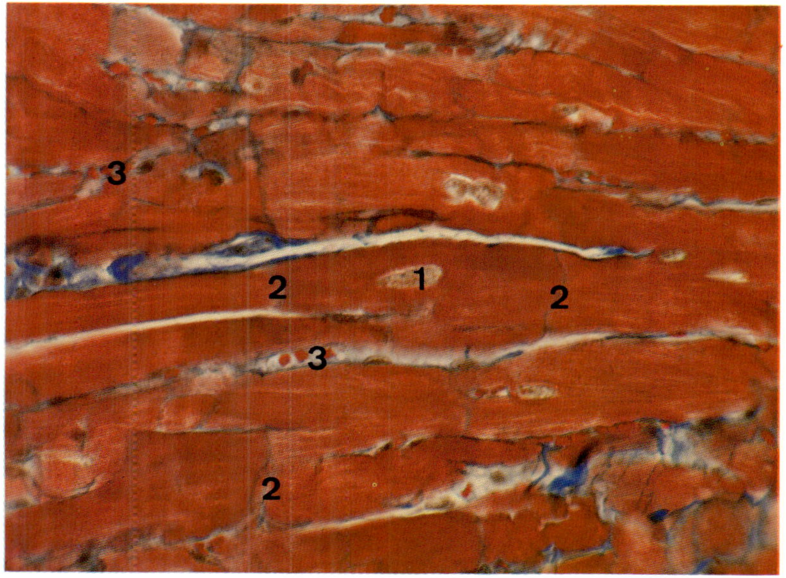

Figure 153

154. Pulmonary vein — Phosphotungstic acid-hematoxylin — 50×

Part of a valve is seen extending across the lumen of the vein.

1. Valve leaflet
2. Alveolus of lung
3. Blood in lumen of vein

155. Cardiac muscle in pulmonary vein — Heidenhain's Azan — 500×

The spherical structure at upper right is a cross section of a nerve trunk. Cardiac muscle of the pulmonary vein forms an arc at the bottom of the field, below which are blood cells in the lumen. Broad elastic fibers stained light blue form part of the tunica media of a large artery at upper left, blood cells filling the arterial lumen.

1. Cardiac muscle
2. Elastic fibers—smooth muscle
3. Blood in artery and vein
4. Nerve trunk
5. Blue collagen fibers

156. Pulmonary vein — Phosphotungstic acid-hematoxylin — 500×

Elastic fibers in the internal elastic membrane and tunica media characterize the large arterial wall in the upper field. Just below are prominent cross striations of cardiac muscle, forming the wall of the pulmonary vein. Blood cells fill the lumina of both artery and vein.

1. Blood cells in lumen of artery
2. Elastic fibers and smooth muscle in arterial wall
3. Cross striations of cardiac muscle in wall of pulmonary vein
4. Blood cells in lumen of vein

157. Heart valves — H & E — 75×

The valve and valve leaflets are located in the upper left of the field, below which are blood cells within the aorta. Myocardium occupies the central field, and another valve lies in the lumen of the pulmonary vein at upper right.

1. Valve
2. Valve leaflet
3. Cardiac muscle
4. Blood in heart chamber
5. Aorta

VETERINARY FIELD STATION LIBRARY

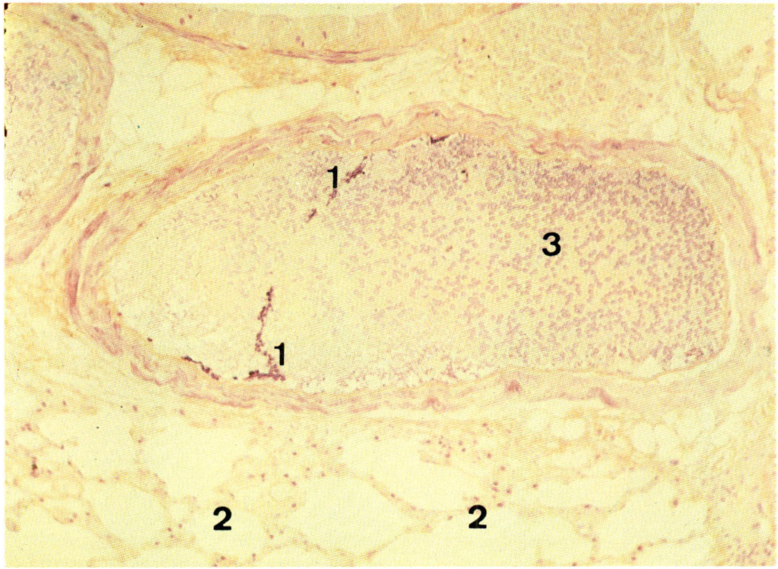

Figure 154

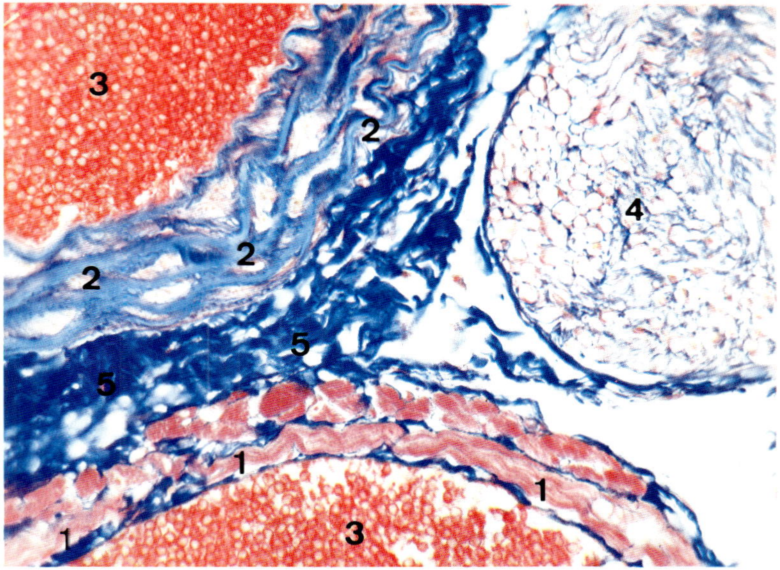

Figure 155

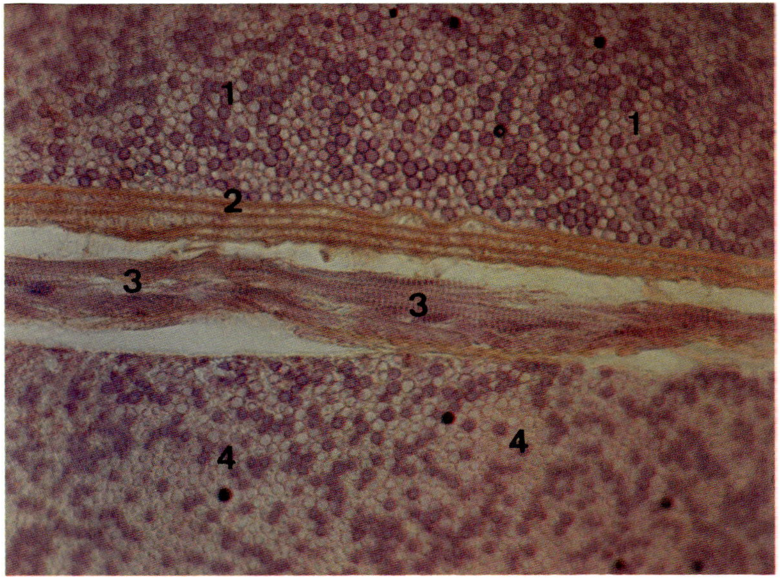

Figure 156

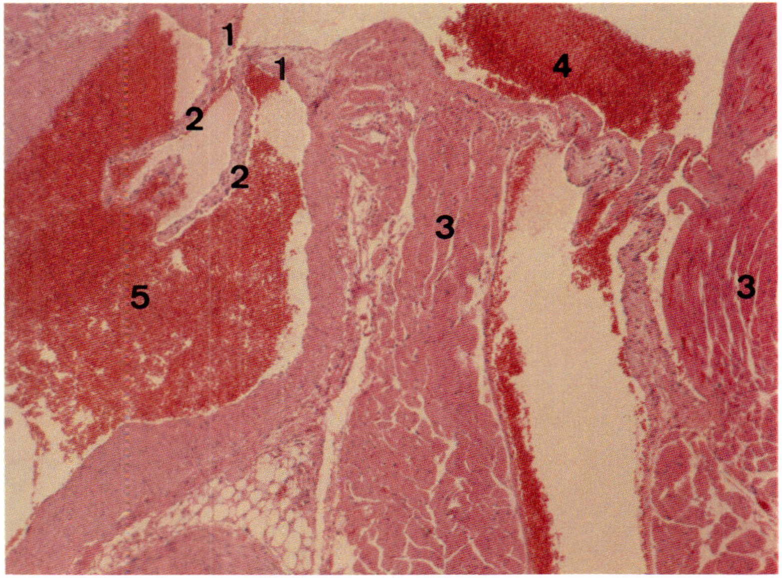

Figure 157

158. Bone marrow Giemsa stain 500×

Many primitive, maturing, and mature cells are found in bone marrow. Cells of the myelocyte and erythrocyte series are present, as are plasma cells and megakaryocytes.

1. Plasma cell
2. Monocyte
3. Neutrophil
4. Normoblast
5. Neutrophil myelocyte

159. Bone marrow Giemsa stain 500×

Two megakaryocytes are seen in the lower right of the field.

1. Megakaryocyte
2. Plasma cell
3. Normoblast
4. Neutrophil

160. Peripheral blood New Methylene Blue 500×

Reticulocytes with a pronounced reticular network are scattered among erythrocytes.

1. Reticulocyte (blue reticular net)
2. Erythrocyte

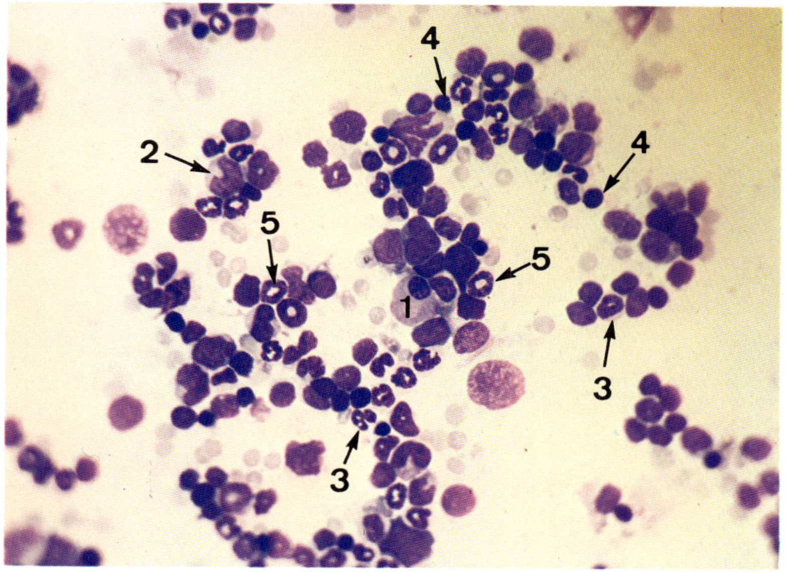

Figure 158

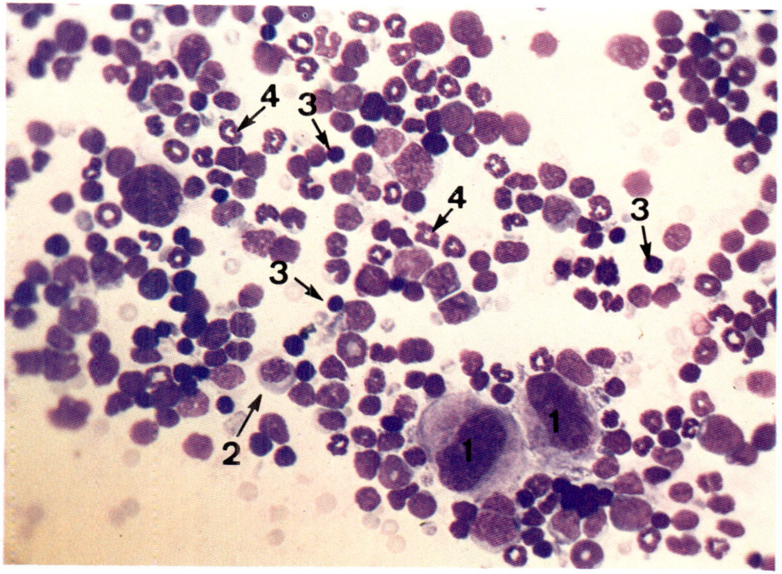

Figure 159

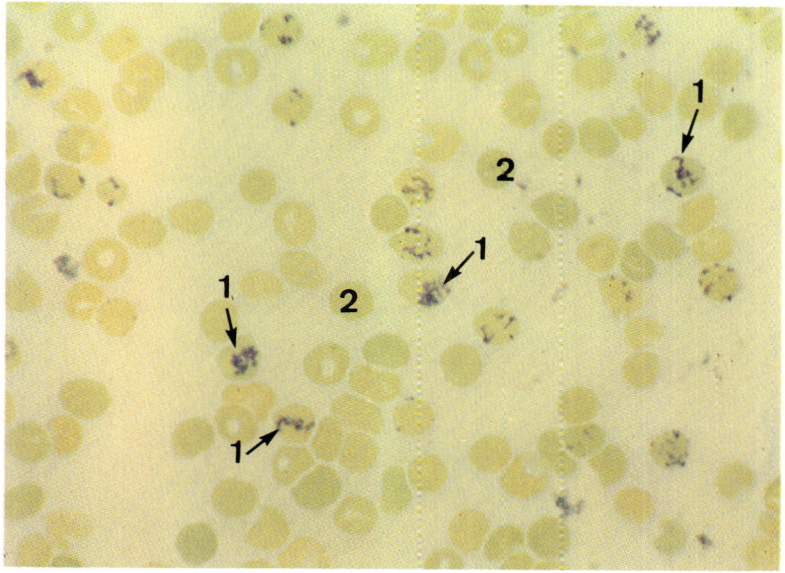

Figure 160

161. Monocyte Wright's stain 950×

This cell is easily identified by the indented, horseshoe-shaped, reddish purple nucleus with light blue cytoplasm. Concave disks of uniform size are erythrocytes.

162. Neutrophils Wright's stain 950×

Both cells exhibit small, bluish cytoplasmic granules, but the nuclear shapes differ. The "doughnut"-shaped cell suggests a nucleus that has not yet formed three lobes, as are seen in the other cell.

163. Platelets Wright's stain 950×

Red granules and blue cytoplasm are readily differentiated in this cluster of platelets.

164. Lymphocytes Wright's stain 950×

Comparison between large and small lymphocytes can be made, although each cell has a large, reddish purple nucleus occupying most of the cell, with a minimum of blue cytoplasm. A clump of platelets with red granules and blue cytoplasm lies between the two cells.

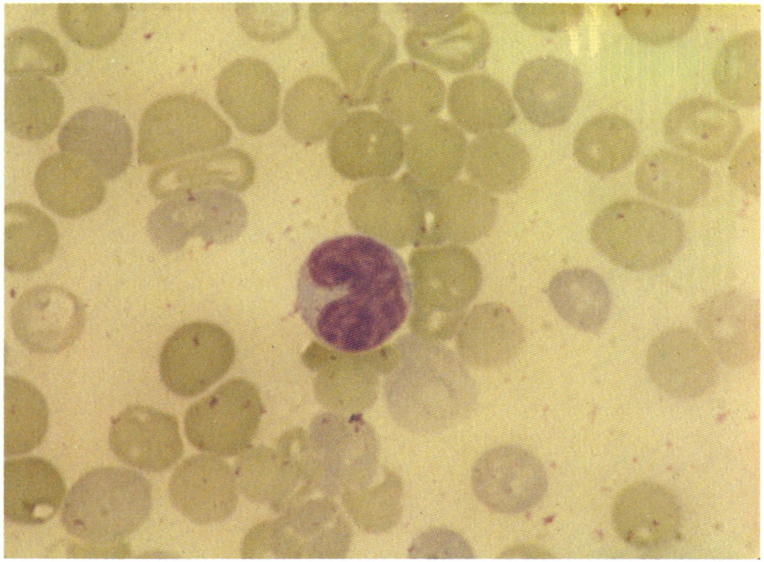
Figure 161

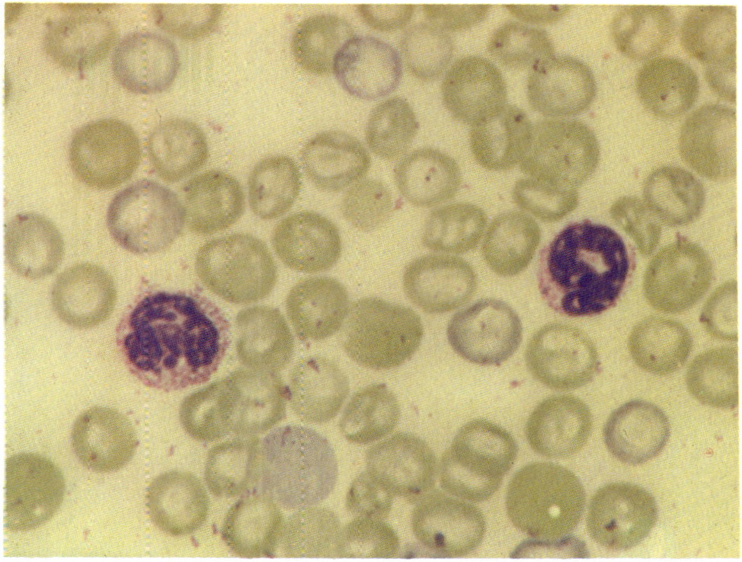
Figure 162

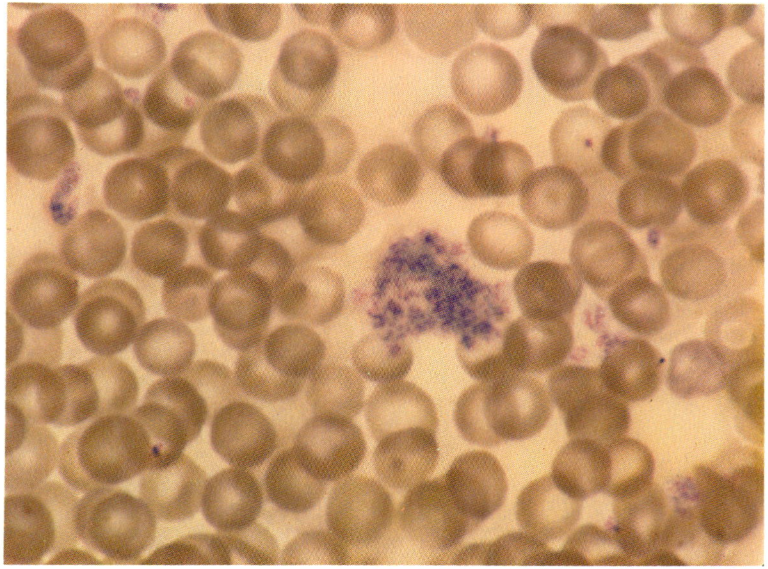
Figure 163

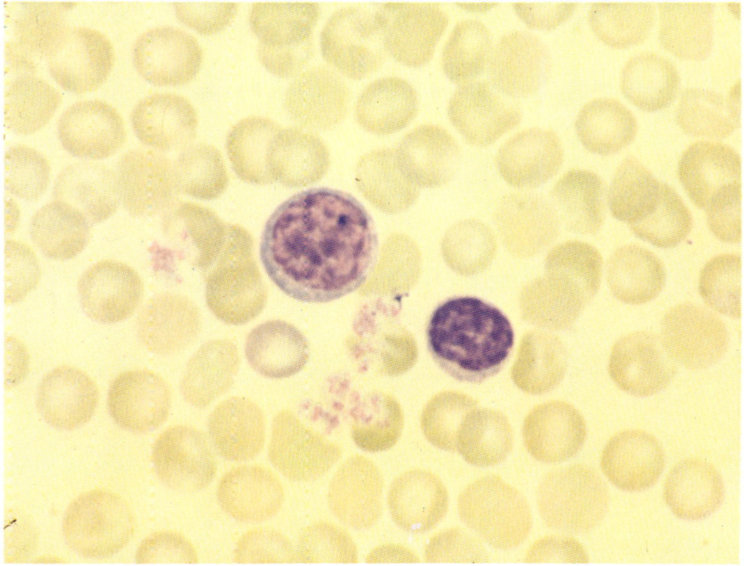
Figure 164

165–168. Eosinophils Wright's stain 950×

These four cells illustrate variations of the nuclear shape and the size and staining intensity of large cytoplasmic granules in eosinophils.

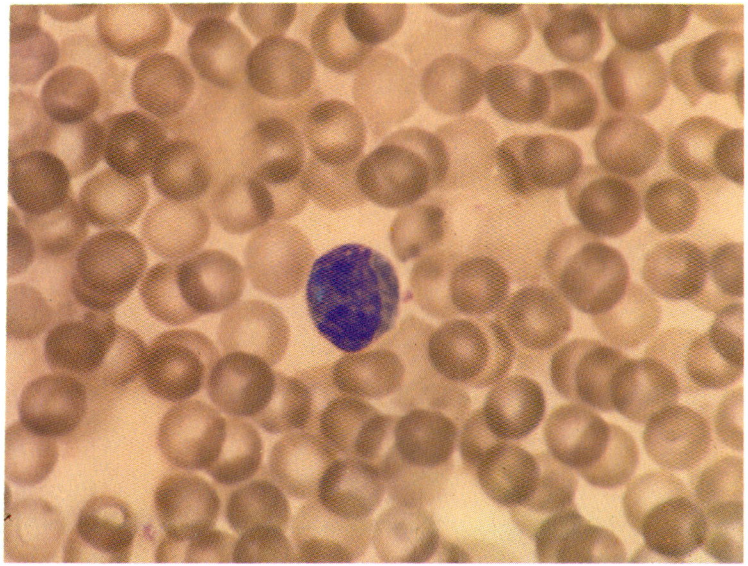

Figure 165

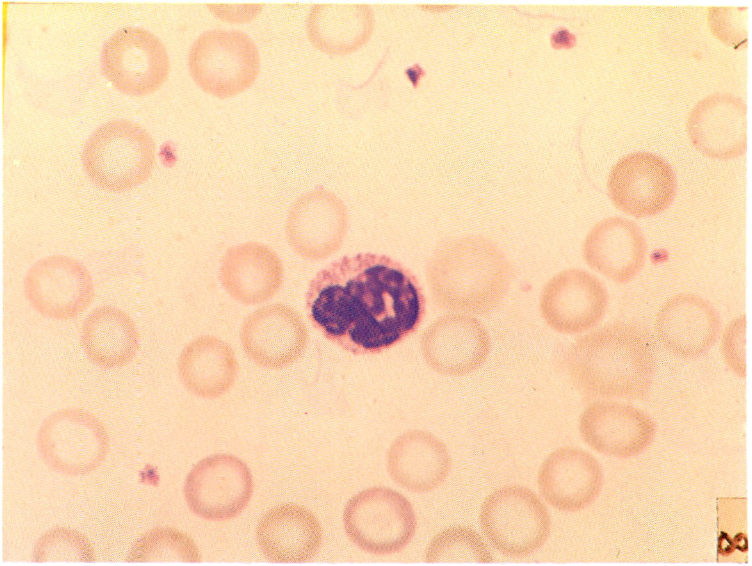

Figure 166

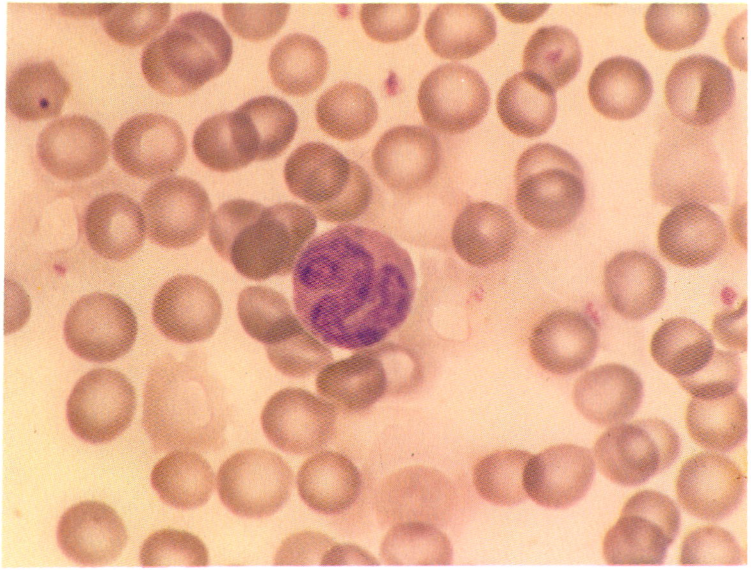

Figure 167

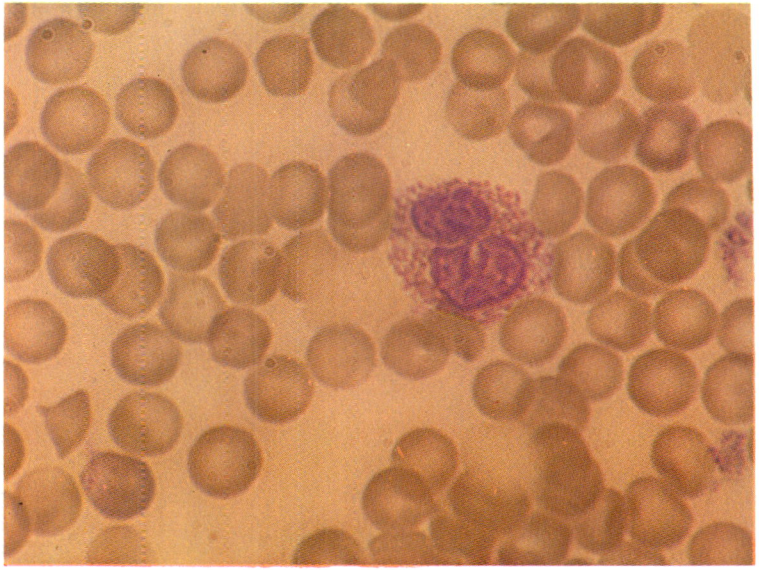

Figure 168

Fixatives

10% NEUTRAL BUFFERED FORMALIN (pH 7.0)

Formaldehyde (37–40%)	100 ml
Distilled water	900 ml
Dibasic sodium phosphate (Na_2HPO_4)	6.5 g
Monobasic sodium phosphate (NaH_2PO_4)	4 g

BOUIN'S FIXATIVE

Picric acid, saturated aqueous solution	750 ml
Formaldehyde (37–40%)	250 ml
Glacial acetic acid	50 ml

Fix 8–24 hr. Store tissues in 70% ethanol. Picric acid is usually removed when sections pass through 70% and 50% ethanol, 10 min each.

ZENKER-FORMOL (HELLY'S FIXATIVE)

Distilled water	1000 ml
Mercuric chloride	50 g
Dissolve with heat	
Add potassium dichromate	25 g

Just before adding tissues add 5 ml 35–40% formaldehyde per 100 ml fixative. Fix 8–24 hr. Wash 24 hr in running water. Store in 70% ethanol. Mercuric chloride must be removed from tissue on slides by alcohol–iodine when staining. For pieces of liver and spleen 2–4 mm thick, 8- to 10-hr fixation is sufficient.

Excellent for H & E and connective tissue stains. Presence of mercuric chloride may interfere with silver staining. Recommend 10% formalin.

CARNOY'S FIXATIVE

Absolute ethanol	60 ml
Chloroform	30 ml
Glacial acetic acid	10 ml

Very fast in penetrating tissues. Excellent for preserving nucleic acids (DNA and RNA).

WILLIAMS'S FIXATIVE FOR BODIAN STAIN

Formaldehyde (37–40%)	40 ml
Glacial acetic acid	10 ml
80% Ethanol	100 ml
Picric acid	2 g

FORMOL-SALINE FIXATIVE

Formaldehyde (37–40%)	100 ml
Sodium chloride	9 g
Distilled water	900 ml

FORMALIN-MERCURIC CHLORIDE

10% Neutral buffered formalin	100 ml
Mercuric chloride	2 g

Dissolve with aid of heat

FORMOL-ALCOHOL-ACETIC (FAA)

Formaldehyde (37–40%)	10 ml
80% Ethanol	90 ml
Glacial acetic acid	5 ml

Used as a rapid fixer.

VANDERGRIFF'S FIXATIVE

95% Ethanol (or 99% isopropyl)	80 ml
Formaldehyde (37–40%)	12 ml
Glacial acetic acid	4.5 ml
Picric acid	4.0 g
Mercuric chloride	0.2 g
Urea	0.5 g

Staining Methods

There are as many ways of preparing tissues for microscopic examination as there are technicians preparing them. The methods we have used and from which our photomicrographs were made are slight modifications from standard procedures given in the references. We have included them in detail in this book to aid the investigator or technician.

DEPARAFFINIZING SECTIONS

The standard procedure for deparaffinizing sections in our laboratory is as follows:

Paraway (Technicon Instruments Corp., Tarrytown, New York 10591) or Xylene	5 min
Xylene	5 min
Drain excess xylene on paper towel before going into absolute ethanol.	
Absolute ethanol	Two changes: First rinse 8–10 times; second, 2–3 min
95% Ethanol	Rinse 8–10 times
80% Ethanol plus iodine	5 min
4% Tincture of iodine	10 ml
80% Ethanol	95 ml
(Used only for tissues fixed in solutions containing mercuric chloride)	
70% Ethanol*	5 min
Tap water	Rinse

Desired specific staining method may now be applied.

HEMATOXYLIN AND EOSIN

Fixative: Any general fixative. Zenker–formol recommended.

Harris's hematoxylin	3–5 min
Hematoxylin crystals	5 g
Absolute ethanol	50 ml
Ammonium or potassium alum (aluminum ammonium sulfate or aluminum potassium sulfate)	100 g
Distilled water	1000 ml
Mercuric oxide (red)	2.5 g

Dissolve hematoxylin in absolute ethanol.

Heat approximately 900 ml of water and dissolve alum. Bring to boil rapidly. Remove from heat; place flask in sink in case contents overflow when adding mercuric oxide. *Reheat briefly* to bring to purple color; cool in running cold water. For use add 1 ml glacial acetic acid for each 100 ml of stain. Filter before use if film forms on surface. Rinse in water 8–10 times.

*Tissues may be left in this solution overnight.

Lithium carbonate, saturated aqueous solution 5–10 sec
(Used to "blue" hematoxylin in nuclei; pH of $LiCO_3$ solution basic to counteract slightly acid hematoxylin)
Rinse in water.
Rinse in 70% ethanol.
Eosin,[1] stock solution

Eosin Y, alcohol and water solution	5 g
Distilled water	100 ml
Add 95% ethanol.	400 ml

Working solution 2–3 min
Eosin stock solution, 1 part
80% ethanol, 3 parts
Add 0.5 glacial acetic acid per 100 ml stain.
Rinse in 95% ethanol. 2 changes
Formalin-fixed tissues may lose eosin more readily than other fixatives.
Suggest only one change.
Rinse in absolute ethanol.
Carbol-xylol 1–2 min

Phenol, fused, crystals melted (J. T. Baker, USP grade)	20 ml
Xylene	80 ml

(Removes any residual moisture present in alcohols)
Xylene 2 changes, 5 min each
Mount in Permount (Fisher), Coverbond (Harleco, Gibbstown, New Jersey) or other suitable medium.

Results: Nuclei: Dark blue
Cytoplasm: Orange red

STAIN FOR MOTOR END PLATES[2]

A. Place frozen sections 20–25 μm thick of skeletal muscle (gastrocnemius, solenus, or tibialis anterior recommended) on a slide in a drop of 3% disodium EDTA (used to prevent contraction of muscle fibers). Let air dry at room temperature.
B. Immerse slides in 20% aqueous solution of sodium sulfate for 3 min to prevent shrinkage of sections in cholinesterase stain. Rinse in distilled water.
C. Stain sections for acetylcholinesterase by incubating at 37°C for 10–25 min in the following solution:

5-Bromoindoxyl acetate* (dissolve first in absolute ethanol)	4 mg
Absolute ethanol	0.3 ml
Potassium ferrocyanide	63 mg
Potassium ferricyanide	50 mg
Tris-HCl (Sigma Chemicals)	42 mg
Tris-base (Sigma Chemicals)	4 mg
Calcium chloride	33 mg
Deionized water	30 ml

Continue staining until microscopy shows end plates to be clearly defined transparent blue. The Tris (mixed to pH 7.2), calcium chloride, and water can be premixed and kept as stock solution.

*Available from ICN Pharmaceuticals, Inc., Life Sciences Group, Plainview, New York.

Nerve-Staining Procedure

1. To prevent loss of the sections from slides, dehydrate in 70% and 100% ethanol for 1–2 min each and then immerse in 0.5–1.0% (w/v) celloidin in 1:1 ether–ethanol solution for 20 sec. Wipe excess from back of slide and harden in 70% ethanol for 30 sec.
2. Fix sections for 30 min at room temperature in buffered formol-saline (37–40% formaldehyde, 20 ml; sodium chloride, 4.25 g; acid sodium phosphate monohydrate, 0.80 g; anhydrous disodium phosphate, 1.30 g; deionized water, 180 ml) at pH 7.0. The fixing solution must be less than one week old.
3. Pretreat for 30 min at 37°C in 10% chloral hydrate (w/v) with 1% pyridine (v/v). Chloral hydrate alone results in distinct staining of terminal nerve branches, but it has the unwanted effect of darkening muscle nuclei and connective tissue. The addition of pyridine suppresses nuclear staining and increases staining of the more proximal axons.
4. Incubate for 40 min at 37°C in 20% silver nitrate containing 0.1% cupric sulfate, with 100 mg calcium carbonate at the bottom of the staining jar. This solution should be freshly made each day.
5. Develop in a solution of 1% hydroquinone and 5% sodium sulfite. Use two baths, the first for 10 sec and the second for 3–5 min. Discard solutions when cloudy.
6. Fix for 2–5 min in 5% sodium thiosulfate.
7. Tone for 5 min in 0.2% sodium tetrachloroaurate (gold chloride) containing one drop of glacial acetic acid per 100 ml. This solution may be used several times if the edges and backs of slides are cleaned before immersion. However, it must be discarded if a precipitate forms.
8. *Optional:* To darken axons and endings, immerse in 1% oxalic acid for 20–120 sec.
9. Immediately fix again for 5 min in 5% sodium thiosulfate.
10. Dehydrate through 70% and 100% ethanol. Remove celloidin by immersion in 1:1 ether–ethanol solution for 5 min. Rinse in xylene and mount.

The reaction may be inhibited by placing the slides immediately after step B in DFP (diisopropyl flurophosphate). CAUTION: DFP is an extremely *poisonous* compound. It should be handled only by a person wearing surgical gloves working under a well-ventilated hood.

For use: approximately 0.1 g DFP in 500 ml distilled water. Treat sections for 30 min at room temperature. Then rinse with two changes of distilled water and proceed with step C.

PERIODIC ACID–SCHIFF STAIN (PAS)

0.5% Aqueous solution periodic acid — 10 min

Rinse in distilled water.

Schiff reagent[3]: — 30 min

Dissolve 1 g basic fuchsin in 200 ml boiling distilled water. Use large (1000 ml) Pyrex Ehrlenmeyer flask. Solution may boil over when dye is added, so flask should be placed in a sink. Shake vigorously to dissolve dye and cool in running water to 50°C. Add 20 ml normal HCl. Cool to 25°C and add 1 g of potassium metabisulfite ($Na_2S_2O_5$). Allow solution to stand in dark 18–24 hr. Add 2 g activated neutral charcoal and shake thoroughly. Filter. Solution should be water-clear. Store in brown bottle in refrigerator.

Rinse in 1% aqueous potassium metabisulfite.
Wash in running water 5–10 min (red color is enhanced).
Counterstain in Harris's hematoxylin. 2–3 min
Rinse in tap water.
Lithium carbonate saturated aqueous solution 2–3 sec
Rinse in distilled water.
Dehydrate in 95% ethanol.
Rinse in absolute ethanol. 2 changes
Clear in xylene and mount.

LUXOL FAST BLUE[4]–NUCLEAR FAST RED

Fix in 10% formalin or Williams's fixative.
Deparaffinize to 95% ethanol.
Stain in Luxol Fast Blue at 57–60°C overnight.

Luxol Fast Blue MSBN (E. I. DuPont de Nemours Co., Wilmington, Delaware)	0.1 g
95% Ethanol	100 ml
10% Aqueous acetic acid	0.5 ml

Rinse in 2 changes of 95% ethanol to remove excess stain.
Wash in distilled water. 2–3 min
Dip in lithium carbonate. 3–4 sec

$LiCO_3$ 1% aqueous solution	5 ml
Distilled water*	95 ml

Rinse rapidly in 2 changes distilled water.
Stain in Nuclear Fast Red solution. 2–5 min

Nuclear Fast Red (Kernechtrot) (Roboz Surgical Instrument Co., Washington, D.C.)	0.1 g
5% aluminum sulfate (5 g/100 ml water)	100 ml

Dissolve with heat, cool, and filter.
Can be used for 3–4 wk. Staining intense in freshly made stain. Reaction is reduced when solution separates into dark and light red areas.
Rinse in 2 changes of distilled water.
Dehydrate through 95% ethanol, then absolute ethanol (2 changes).
Clear in xylene and mount.

Results: Myelinated nerve fibers: Greenish-blue
Nuclei of neurons and neuroglial cells: Bright red

ALCIAN BLUE–TARTRAZINE STAIN[6]

Fix tissues 24 hr in 10% formal saline.
Deparaffinize and bring sections to water.
Oxidize 5 min in freshly prepared equal volume mixture of 0.3% aqueous potassium permanganate and 0.3% aqueous sulfuric acid. Bleach in 3% aqueous sodium metabisulfite. Wash in running tap water 5 min.

*This step in differentiation is important: White matter of nerve fibers is greenish blue. Gray matter (nerve cell bodies) is colorless.

Stain in celestine blue.	3–5 min
Celestine blue (Roboz Surgical Instrument Co., Washington, D.C.)	0.25 g
Ferric ammonium sulfate (iron alum) (J. T. Baker Co., Phillipsburg, New Jersey), 5 g/100 ml water	50 ml
Glycerine	7 ml
Rinse in water.	
Harris's or Mayer's hematoxylin	2–3 min
Blue in running water.	5–10 min
Stain in Alcian Blue solution.	30 min
Alcian Blue 8 GS or 8 GX (Roboz Surgical Instrument Co., Washington, D.C.)	1 g
Distilled water	100 ml
Rinse in distilled water.	
Wash in running water.	5 min
Stain in tartrazine solution.	30 sec
Tartrazine (Roboz Surgical Instrument Co., Washington, D.C.)	2 g
1% Aqueous acetic acid	100 ml
Rinse in distilled water.	
Dehydrate rapidly through	
95% Ethanol	
Absolute ethanol	2 changes
Clear in xylene and mount.	

Results: Nuclei: Blue-black
Epithelial lining: Green
Mucous neck cells: Pale green or bluish
Zymogenic granules of chief cells: Dark green
Parietal cells: Yellow
Erythrocytes: Bright yellow

Mayer's hematoxylin	
Hematoxylin	1 g
Distilled water	1000 ml
Ammonium or potassium alum	50 g
Sodium iodate	0.2 g
Citric acid	1 g
Chloral hydrate	50 g

Dissolve the alum in water, warming gently. Dissolve hematoxylin in this solution. Add citric acid, sodium iodate, and chloral hydrate. Shake well to dissolve all components. Final color of solution is reddish-violet. Keeps well for months.

MASSON CONNECTIVE TISSUE STAIN[1]

Fixative: Any general fixative; Bouin's and Zenker–formol recommended.

Following deparaffinization: stain in celestine blue–hematoxylin (see under Alcian Blue–Tartrazine stain).	
Wash in running water to improve bluing.	
Stain in Biebrich scarlet–acid fuchsin.	5 min
Biebrich scarlet, 1% aqueous solution	90 ml
Acid fuchsin, 1% aqueous solution	10 ml
Glacial acetic acid	1 ml
Rinse in distilled water.	2 changes
Phosphomolybdic–phosphotungstic acid	5 min
Phosphomolybdic acid	5 g
Phosphotungstic acid	5 g
Distilled water	200 ml
(Transfer directly to Fast Green solution.)	
Fast Green FCF, 0.5% aqueous solution	1–2 min

1% Acetic acid	rinse
Distilled water	rinse
Dehydrate through 95% ethanol, absolute ethanol.	2 changes
Clear in xylene and mount.	

Results: Nuclei: Deep mauve-blue
Cytoplasmic elements: Varying shades of red
Muscle: Red
Collagen: Green

MALLORY-HEIDENHAIN'S AZAN STAIN[6]

Fixative: Zenker–formol, Bouin's; mordant formalin-fixed tissues overnight in 3% potassium dichromate to improve staining.

Treat with aniline alcohol.	45 min
Aniline	5–10 drops
95% Ethanol	100 ml
Treat with acid alcohol.	1–2 min
Glacial acetic acid	1 ml
95% Ethanol	100 ml
Rinse in distilled water.	
Stain in Azocarmine at 55–57°C.	1 hr
Azocarmine G	0.5 g
Distilled water	100 ml
Glacial acetic acid	1 ml
Rinse in distilled water.	2 changes
Differentiate in aniline–alcohol.*	

*Nuclei, brilliant red; cytoplasm, very light red. Check with microscope.

Rinse in acid alcohol.	
Rinse in distilled water.	2 changes
Transfer to phosphotungstic acid.	2 hr
Phosphotungstic acid	5 g
Distilled water	100 ml
Rinse in distilled water.	
Stain in Aniline Blue solution.	2 hr
Aniline Blue, water solution	0.5 g
Orange G	2 g
Oxalic acid	2 g
Distilled water	100 ml
5% Aqueous phosphotungstic acid	1 ml
Rinse in distilled water.	
Treat with 5% aqueous solution phosphotungstic acid.	3–5 min
Rinse in distilled water.	
Rinse in aqueous 1% acetic acid.	
Dehydrate through 95% ethanol.	2 changes
Dehydrate through absolute ethanol.	2 changes
Clear in xylene and mount.	

Results: Nuclei: Brilliant red
Collagen and reticulum: Blue
Muscle: Red and yellow
Anterior pituitary cells:
- Basophilic cytoplasm: Blue
- Acidophilic cytoplasm: Orange-red
- Chromophobes: Colorless or light gray

By overstaining with Aniline Blue solution (18–24 hr) striking staining effects may be achieved.

SNOOK'S METHOD FOR RETICULUM[1]

Fixative: Williams's fixative, or 10% neutral buffered formalin, recommended

Oxidize in 0.25% aqueous solution potassium permanganate.	5 min
Rinse in distilled water.	
Oxalic acid 5% aqueous solution (until brown sections are clear)	3–5 min
Wash in running tap water.	2–3 min
Mordant in 1% aqueous uranium nitrate solution.	5–10 sec
Wash in running water.	3–5 min
Ammoniacal silver solution	1 min

To 20 ml of 5% aqueous silver nitrate solution in a 125-ml acid-cleaned Ehrlenmeyer flask add 20 drops of aqueous 10% sodium hydroxide solution. Heavy brown precipitate forms. Dissolve by adding ammonium hydroxide, drop by drop, shaking gently, leaving only a few granules at the bottom of the flask. Add distilled water to make 60 ml and use at once.

Rinse in 2 changes distilled water.	
Develop in 1% aqueous formalin solution.	1 min
Rinse in 2 changes distilled water.	
Tone in 1% aqueous gold chloride solution until section is grayish black.	1–2 min
Rinse in 2 changes distilled water.	
Sodium thiosulfate, 5% aqueous solution	5 min
Rinse in 2 changes distilled water.	
Counterstain in Nuclear Fast Red (see Luxol Fast Blue–Nuclear Fast Red Stain).	2–3 min
Rinse in 2 changes distilled water.	
Dehydrate through 95% ethanol.	2 changes
Dehydrate through absolute ethanol.	2 changes
Clear in xylene.	2 changes, 5 min each
Mount in Permount or suitable mounting medium.	

Results: Reticular fibers: Gray to black
Nuclei: Red

BODIAN METHOD FOR NERVE FIBERS[6]

Fix overnight in Williams's fixative or 10% neutral buffered formalin.

Formaldehyde 35–40%	40 ml
Glacial acetic acid	10 ml
80% Ethanol	100 ml
Picric acid	2 g

Place tissues in and store in

70% ethanol, 2 changes, to remove most of picric acid	10–20 min

Stain paraffin sections in

Strong silver protein (Roboz Surgical Instrument Co., Washington, D.C.)	1 g
Distilled water	100 ml

Place slides in staining dish containing water; distribute 5–6 g clean copper (wire, shot, or sheet) between slides and around bottom of dish. *Sprinkle* silver protein on top of water evenly and place staining dish in oven at 37°C for 12–24 hr. (Do not dissolve by stirring.) Cover with glass top.

Distilled water (2 changes), rinse	8–10 times

Place in reducing solution.	5–10 min
Hydroquinone	1 g
Sodium sulfite	5 g
Distilled water	100 ml
Rinse in 3 changes of water.	1 min total
Tone in	1–2 min
Gold chloride	1 g
Distilled water	100 ml
Rinse in 3 changes of water.	1 min total
Develop in 2% aqueous oxalic acid.	2–4 min

Examine slide with microscope. Fibers become dark brown, then gray to black.

Wash in distilled water, 3 changes.	1 min each
Fix in 5% aqueous sodium thiosulfate.	5 min
Wash in running water.	5–10 min

Rinse in distilled water.

Dehydrate through 95% ethanol, 2 changes absolute ethanol. Clear in xylene (2 changes, 5 min each) and mount.

Results: Nerve fibers: Dark brown to black

METHACRYLATE SECTIONS

Lee's Methylene Blue–Basic Fuchsin Stain[7]:

This general histological stain is used to differentiate cell types in the pituitary and endocrine cells in the intestinal and stomach epithelium. It displays C cells of the thyroid and clearly differentiates between smooth muscle and collagen. Elastic fibers and mitochondria are clearly revealed, and juxtaglomerular granules and mitochondria in the kidney stain bright red.

Stock solutions:

1. 0.13 g methylene blue in 100 ml deionized water
2. 0.13 basic fuchsin in 100 ml deionized water
3. 0.2 M phosphate buffer, pH 7.2–7.8

Staining solution: Mix

Methylene blue solution	12 ml
Basic fuchsin solution	12 ml
0.2 M phosphate buffer	21 ml
Ethanol (95% or absolute)	15 ml

This filtered solution is useful for 4 or 5 days.

Staining procedure: The sections are immersed in the staining solution for 10–15 sec.* The slides are rinsed briefly in deionized water and allowed to dry. The dried slides are mounted under a cover slip.

Phosphate buffer:

A. Sodium phosphate, monobasic ($NaH_2PO_4 \cdot H_2O$)	13.8 g
Distilled water	500 ml
B. Sodium phosphate, dibasic (Na_2HPO_4)	26.8 g
Distilled water	

For pH 7.2, add

Monobasic sodium phosphate	10 ml
Dibasic sodium phosphate	40 ml

*It has been our experience that this is too short a time for various tissues. By staining a test slide for about 1 min the nuclear components should take the blue stain. By allowing the slide to stay in the stain 1–2 min more, the basic fuchsin should stain the connective tissue and erythrocytes brilliant red.

PROCEDURE FOR PREPARING METHACRYLATE SECTIONS

Reagents from JB-4 Embedding Kit (Polysciences, Inc., Paul Valley Industrial Park, Warrington, Pennsylvania 18976)

Fixation: Neutral buffered formalin at room temperature 4–24 hr depending upon size

Distilled water	900 ml
Formaldehyde, 35–40%	100 ml
Na_2HPO_4	6.5 g
$NaH_2PO_4 \cdot H_2O$	4 g

One lobe of a mouse lung (1.25 cm × 0.63 cm thick) can be fixed in 4 hr.

If tissues cannot be processed immediately after fixation, store in 70% ethanol.

Other common fixatives may be used, except glutaraldehyde.

Dehydration: 70% Ethanol for 30 min
80% Ethanol for 30 min
95% Ethanol for 30 min
Absolute ethanol, 2 changes, 30 min each

Catalyzed Solution A 2 hr, under vacuum, at room temperature

Embedding:

Catalyzed Solution A	25 parts
Solution B	1 part

Mix well, pour into mold, and orient tissue.
Should harden within 1 hr, ready to section.

Cutting: Sections routinely cut at 1.5 μm using glass knife with JB-4 microtome (Sorvall)

Staining: See Lee's Methylene Blue–Basic Fuchsin stain.

ALDEHYDE FUCHSIN TRICHROME STAIN[8]

Fixative: Zenker–formol; Bouin's

Deparaffinize through xylenes.
Hydrate through absolute and 95% alcohols.

80% Ethanol plus iodine	5 min
70% Ethanol	5 min
Water	Rinse
Oxidize in 0.3% potassium permanganate.	2–3 min

Freshly prepare by mixing equal parts of 0.6% $KMNO_4$ and 0.6% H_2SO_4.

Rinse in distilled water.

Decolorize in 4% aqueous sodium bisulfite ($NaSO_3$).	1–3 min
Wash in running water.	5 min
Stain in aldehyde–fuchsin solution.	1–3 min

Add 3 ml paraldehyde and 1 ml concentrated HCl to 0.3% solution of basic fuchsin in 60% ethanol.

Staining solution should mature for 3 days before use. Discard after 30 days.

Rinse in 95% ethanol.	3 changes
Wash in running water until background is pale.	1 hr

Stain in celestine blue–hematoxylin (see Alcian Blue–Tartrazine Stain).

Rinse in water.
"Blue" in lithium carbonate. 5–10 sec
Rinse in water.
Stain in Ponceau–fuchsin mixture. 2 hr

1% Ponceau–xylidine	4 parts
1% Acid fuchsin	1 part
1% Fuchsin S	1 part

Dilute 1:10 in 1% acetic acid.
Rinse in 1% acetic acid.
Rinse in distilled water.
Stain in 1% solution of light green in 1% acetic acid. 5–15 min
Rinse in distilled water.
Differentiate in 0.5% phosphomolybdic acid. 15 min
Place in 1% acetic acid. 20 min
Dehydrate rapidly through absolute ethanol. 3 changes
Clear in xylene and mount.

Results: B cell granules: Purple
A cells: Red
Elastic tissue and mucus: Purple
Collagen: Green

PHOSPHOTUNGSTIC ACID–HEMATOXYLIN (PTAH)

Fixative: Zenker or any fixative with mercuric chloride recommended.

Following deparaffinization, down to water through graded alcohols, tissues are oxidized in freshly prepared 0.5% potassium permanganate. 5 min
Rinse in water.
Bleach in 5% oxalic acid. 5 min
Wash in running water. 5 min
Mordant in 4% iron alum. 1 hr
Wash in running water. 10 min
Stain in PTAH (4 hr usually is sufficient). 3–24 hr

Distilled water	1000 ml
Hematoxylin	1 g
Phosphotungstic acid	20 g
Add potassium permanganate to aid immediate ripening	0.18 g

Transfer sections directly to 95% ethanol for brief rinse.
Absolute ethanol, rapid rinse
Clear in xylene.
Mount in Permount or suitable mounting medium.

Results: Nuclei, neuroglial fibers, contractile elements of striated muscle, fibrin: Blue
Collagen, elastin, cartilage, bone ground substance: Yellowish to brownish red

WRIGHT'S STAIN

Place smears on staining rack and stain 2 min with stock Wright's stain.

Wright's stain, dry	0.1 g
Absolute methanol	60 ml

The dry stain may be mixed with the alcohol by grinding, using a mortar and pestle, or, to avoid well-stained fingers, the alcohol and dry stain may be mixed in a suitable flask and placed on an automatic shaker for 4–6 days. A satisfactory solution is obtained.

Flood the slide with distilled water, thus diluting the staining solution. — 4 min

Buffered diluting solution may be necessary if the distilled water is too acid.

Phosphate buffer:	Disodium phosphate (Na_2HPO_4)	7.1 g
	25% Methanol in distilled water	250 ml
Citric acid:	Citric acid	5.25 g
	25% Methanol in distilled water	250 ml

Test stained smear under microscope. If too red add 3–4 drops phosphate buffer per ml of diluting fluid.

Lymphocytes should have reddish-purple nucleus and light blue cytoplasm. Eosinophils should exhibit brick red cytoplasmic granules and dark, reddish-purple nucleus.

Apply cover glass with mounting medium for permanent preparation.

MODIFIED MARTIN–MALLORY STAIN FOR MOUSE PITUITARY GLAND

Fix in Zenker–formol (95% Zenker fluid, 5% formaldehyde) overnight.

Wash tissue in running tap water 8–10 hr.

For pituitary gland *in situ* decalcify in 10% formic acid. — 2–3 days

Wash in running tap water to remove acid. — 24 hr

Dehydrate and embed in Paraplast or similar embedding medium.

Following deparaffinizing, sections are stained in Harris's hematoxylin — 2–3 min

Rinse in tap water.

"Blue" in saturated aqueous $LiCO_3$. — 1 min

Rinse in water.

Stain in 0.5% aqueous acid fuchsin. — 3–5 min

Differentiate in tap water; acidophil granules should be bright red.

Place in 1% aqueous phosphotungstic acid. — 3–5 min

Pass slides directly into Mallory's Aniline Blue Orange G stain. — 1–2 hr

Aniline Blue, water soluble	0.5 g
Orange G	2.0 g
Phosphotungstic acid, 1% aqueous solution	100 ml

Differentiate in 95% ethanol.

Blue granules are intense but fuchsin-stained granules are clearly visible.

Rinse in two changes absolute ethanol.

Clear in two changes xylene.

Mount with Permount or suitable mounting medium.

Results: Nuclei: Dark blue
Acidophil granules: Bright red
Basophil granules: Blue
Chromophobes: Gray cytoplasm
Erythrocytes: Orange to red
Bone: Intensely blue

NEW METHYLENE BLUE STAIN FOR RETICULOCYTES AND HEINZ BODIES

Place 2 drops oxalated or heparinized whole blood on a clean glass slide.
Add 3–4 drops of New Methylene Blue to the blood, mix well, and allow to stand 10–15 min.

Distilled water	100 ml
Potassium oxalate	1.6 g
Dye	0.5 g

Rinse with distilled water and apply cover glass for viewing under the microscope *or*
Air dry.
Apply cover glass with Permount or any suitable mounting medium for permanent preparation.
Results: Reticulocytes have distinct, blue reticular network.

TOLUIDINE BLUE STAIN

On a warming plate at 60°C flood epon section with staining solution.	4–5 min
1% aqueous solution toluidine blue	1 part
2.5% aqueous solution sodium carbonate (Na_2CO_3)	20 parts

Rinse in tap water.
Methacrylate sections should be stained vertically 4–5 min in a Coplin jar to prevent residue from accumulating on slide.

GIEMSA STAIN

Smears: Fix in absolute methanol.	15 min
Air dry.	
Stain in working solution.	15–30 min
Giemsa stock*	3 ml
Absolute methanol	3 ml
Distilled water	100 ml
Adjust staining solution to pH 6.8–7.0.	
Buffers: 25% Methanol	250 ml
Disodium phosphate (Na_2HPO_4)	7.09 g
25% Methanol	250 ml
Citric acid	5.25 g

Rinse in tap water
Results: Lymphocytes: Nucleus—dark purple; cytoplasm—sky blue
Eosinophils: Nucleus—reddish purple; cytoplasmic granules—brick red

*Giemsa Blood Stain, Azure B type, Harleco, Gibbstown, New Jersey. Giemsa with Azure A MC/B Manufacturing Chemists, Norwood, Ohio.

VETERINARY FIELD STATION LIBRARY

References

1. Lee G. Luna (ed.), *Manual of Histological Staining Methods of the Armed Forces Institute of Pathology,* 3rd ed., McGraw-Hill, 1968.
2. Alan Pestronk and Daniel B. Drachman, A new stain for quantitative measurement of sprouting at neuromuscular junctions, *Muscle and Nerve* **Jan.-Feb.:** 70–74, 1978.
3. A. G. E. Pearse, *Histochemistry, Theoretical and Applied,* J & A Churchill, London, 1960.
4. Edward Gurr, *Staining, Practical and Theoretical,* Williams and Wilkins, Baltimore, 1962.
5. T. A. Asojo, Alcian blue tartrazine: A simple differential stain for the fundic gland cells of mammalian stomach, *Stain Technology* **50:**55–57, 1975.
6. Gretchen L. Humason, *Animal Tissue Techniques,* 3rd ed., W. H. Freeman, San Francisco, 1972.
7. Polysciences, Inc., JB4 Embedding Kit, May 1976, Warrenton, Pennsylvania 18976, Data Sheet 123.
8. Sydney S. Lazarus and Bruno W. Volk, *The Pancreas in Human and Experimental Diabetes,* Grune and Stratton, New York, 1962.

RECOMMENDED READING

William Bloom and Don W. Fawcett, *A Textbook of Histology,* 9th ed., W. B. Saunders, Philadelphia, 1968.

M. B. L. Craigmyle, *Color Atlas of Histology,* Year Book Medical Publishers, Chicago, 1975.

Earl L. Green (ed.), *Biology of the Laboratory Mouse,* 2nd ed., McGraw-Hill, New York, 1966. Chapter 13, Anatomy, Katharine P. Hummel *et al.*

Index

Lightface numbers are text page numbers; boldface numbers are color figure numbers. Figure A will be found on p. 70.